KINKY CATS, IMMORTAL AMOEBAS, AND NINE-ARMED OCTOPUSES

KINKY CATS, IMMORTAL AMOEBAS, AND NINE-ARMED OCTOPUSES

WEIRD, WILD, AND WONDERFUL BEHAVIORS IN THE ANIMAL WORLD

RAYMOND OBSTFELD

HarperPerennial
A Division of HarperCollins*Publishers*

HarperCollins books may be purchased for educational, business, or sales promotional use. For information, please write to: Special Markets Department, HarperCollins Publishers Inc., 10 East 53rd Street, New York, New York 10022.

FIRST EDITION

Designed by Joseph Rutt

Library of Congress Cataloging-in-Publication Data

Obstfeld, Raymond, 1952–
Kinky cats, immortal amoebas, and nine-armed octopuses : weird, wild, and wonderful behaviors in the animal world / Raymond Obstfeld. — 1st ed.
p. cm.
Includes index.
ISBN 0–06–273419–9
1. Animal behavior. 2. Human behavior. 3. Psychology, Comparative. I. Title.
QL751.026 1997
156—dc21 96-40391

---

97 98 99 00 01 ❖/RRD 10 9 8 7 6 5 4 3 2 1

CONTENTS

WHAT THIS BOOK IS ABOUT

A couple years ago I finished writing a book comparing the ethical teachings of the world's major religions. After six years of inhaling stale library dust and staring glassy-eyed at a glowing computer screen, I had read well over a thousand books and articles and written over fifteen hundred pages. I'd gone from perfect eyesight to needing reading glasses, I'd put on ten pounds, and my reverse layup was a couple steps slower. At that point, I found myself unwilling to read or write one more word on the subject.

Yet, here I am again with another book about ethics. Not about human ethics—at least not directly—but about how animal behavior relates to human notions of ethics. It's a natural extension, really, because human morality is clearer to us when we compare it to the alternative options. And the wide variety of animal behavior makes such a comparative study constantly surprising and amazing. Just about every day during the writing of this book, I turned to my wife and said, "Listen to this," and proceeded to read to her yet another startling example of animal behavior.

But if we are to make a comparison, the first question we

have to ask is, Do animals have ethics? The average person on the street would probably say "no" or "kinda" or "I don't know." But ask them, Do humans have ethics? and this same person would say "yes" or "sometimes" or "they could if they wanted." The main distinction between humans and animals—according to this conventional wisdom—is that humans (1) have the ability to reason and therefore determine what a moral act is and (2) use their free will to choose or not choose that moral act. We are moral animals because we *could* be moral—not necessarily because we actually behave morally. Animals, we are told by those same "conventional wisdom" folks, (1) act purely on instinct and learned behavior, (2) are not capable of analyzing the right or wrong of an action, and (3) lack free will to choose an action based on moral content. Therefore, they cannot be moral.

As Sportin' Life crows in *Porgy and Bess*, "It ain't necessarily so."

Technically, humans are animals. Check the textbooks. Still, we often don't think of ourselves as such. In fact, our whole moral structure is based on the conviction that we are able to "rise above" the behavior of other animals. Fido can't help his unceremonious infatuation with my leg; Fluffy can't help torturing that mouse to death. That's how they were designed, so that's what they must do. They are basically organic machines, hardwired to spend their lives fulfilling desires and impulses they don't understand. We humans, however, believe that we alone are able to consciously override any biological programming and act according to ethical principles we have devised through pure intellect or from divine revelation. Like spiritual MacGyvers, we wad some mental chewing gum into our "biological imperative" circuitry, bypass the intruder alarm, and reroute the controls of our mind and body.

Regardless of how our bodies were originally designed—whether to kill when threatened or rape when lustful—our souls have found a way to reprogram. This is the general theory that is the basis for most religions and philosophies in the world. Whether this reprogramming is even possible or is merely a self-serving illusion is a discussion reserved for other more learned books.

Still, one problem with this popular view is that it reduces the influence of biology to a walk-on supporting role. Regardless of how smart we are, we are still encased in physical bodies, and those bodies, like straight-armed zombies wandering the streets, have certain unconscious marching orders: masticate, defecate, replicate. Nothing's going to interfere with that. As much as our heads might be soaring in the clouds with the angels, our teeth still long to rip into some food, and having eaten we still have to eliminate waste; and most people experience the crazed grip of lust and the obsessive torture of love that leads to children. Humans have gone out of their way to degrade their own bodies as the source of evil. We are prisoners of the flesh, the teachings explain. Our souls or spirits are pure goodness, but they are befouled and contaminated by being encased in these carnal bodies. If it weren't for this flesh and its demands, we would be able to escape this world and live with whatever heavenly creatures we believe in. Trapped in the material world, our spirits weighed down by the flesh as if it were the metal armor of a knight, we slog through this life in a constant battle to control the impulses of the body.

Not animals. They don't experience existential *angst*, debating which choice expresses goodness and which evil. They just do. Which is why animals have come to symbolize the incarnation of these awful desires of humans. Many religions justify their use of animals for food, clothing, labor, and experimentation on the grounds that the animal is inferior. The animal can only do what its programming tells it to do. Because humans can do otherwise, we must be better, more godly, and therefore may do with animals as we see fit. Obviously this is once again "conventional wisdom," not the way everyone feels. But it is not much different than the way women were once viewed (and still are in some places). Men were considered the spiritual beings; they were moral and spent their time thinking of God and heaven. Women were the physical beings; they represented human enslavement to the physical body (after all, men argue, it was Eve who put us in this position). A woman was thought to be unable to be moral—unless she had the guidance of a man. If a man was aroused by seeing a woman, the woman was blamed. Everyone knew that the man would have been moral if the woman, through some sorcery, hadn't inflamed his desire. Sometimes a man would have illicit sex with a woman, then she would be driven from the town or killed for bewitching him to sin.

Thousands of years of suppressing and repressing desires and actions based on those lusty desires has left us shuffling toward the twenty-first century confused about just what is natural behavior for humans. For those who believe a God created all beings, certainly such a God would not have designed our bodies so that all its functions were evil. Our moral codes are reactions to desires and guidelines that try to distinguish between natural and perverse. If we were designed to do something, that must be good, but perversions of that design must be bad. Almost every society endorses monogamy, yet 97 percent of mammals are not "naturally" monogamous (humans aren't classified as monogamous). Most *birds* are monogamous, though, and much of our literature points to the habits of birds, such as the lifelong "marriage" of swans, as examples of how humans should act. "See," we say, "it's natural." Homosexuality is often blasted for not being natural. Yet, it is rampant in the animal kingdom, as is transsexuality (changing sex).

Primitive cultures once looked to animal behavior for role models. They mimicked animals' behavior in building shelter, dressed like their favorite animals, and adopted them as spiritual guides. They worshipped them as gods. And as gods, animals were expected to show the way, to teach humans how to act "naturally." But there are also dangers to using animal behavior as a guide for human behavior—especially when such behavior is misinterpreted. This was the case with the male hornbill bird, which walls up its pregnant mate in the hollow of a tree (see MARRIAGE: HORNBILL). Out of respect for the natural ways, some African tribes copied the hornbill's method, though completely misunderstanding its purpose. While the female hornbill was temporarily walled in because she had molted during pregnancy and was therefore completely helpless until her feathers grew back, a wife in one of these tribes would be imprisoned throughout her entire life. Another misunderstanding came when humans noticed that a bird that had eaten a Spanish fly (which is really a beetle with a heart-shaped head and iridescent emerald body) would immediately begin to writhe on the ground in what appeared to be a sexually suggestive manner. Furthermore, since beetles sometimes mate up to twenty hours at a pace so fast it's a blur to watch, humans made the connection between eating this beetle and sexual prowess. However, the suggestive pelvic thrusting of birds—as well as that of humans and other animals after eating one of these beetles—is the result of cantharides in their bodies, which cause an inflamed bladder or urethra. What looks like passion is merely pain.

We still can see remnants of animal worship in the names

we have for our cars (Cobra, Thunderbird, Cougar) and sports teams (Tigers, Bears, Dolphins), and in our national symbol (eagle). Once animals were replaced as gods by humans born of gods, they became little more than quaint icons. Consequently, the more technologically based human society becomes, the more we feel removed from what is natural. The average person often fears that the technology is beyond him and therefore he is intellectually powerless, and that his own animal self has been buried within him and he is therefore instinctually powerless. The rise in popularity of spiritual movements and literature is the direct result of these feelings of alienation and impotence. Whenever we feel this way, we return our gaze to the animal world and once again scrutinize its behavior to better understand our own. We're like trackers checking urine smells on branches in order to find the unicorn we know is just up ahead.

The problem is that when we see animals do something we like, we call it "natural"; when they do something we don't like we call it unnatural. Then we denounce their behavior as "animal" and start waving the humanity flag and singing our anthem, "We Aren't Really Animals, We Just Look That Way." Partially, this dilemma over what is natural exists because we are so removed from our origins. Human beings seem to have a unique ability to learn. We are hungry for information, devouring it endlessly. But we aren't especially discriminating in our intake. We gulp anything that comes our way. Other animals are more discriminating; they learn only what benefits them immediately, what furthers their drive for food and procreation. But humans continue to gather all kinds of information, some immediately useful, some the purpose of which is not immediately apparent. All that information draws us further and further away from whatever we might call instinctual or "animal" behavior. It's like the typical nightmare that most actors share:

They are in a play, on stage (often naked), and don't know what their lines are, what the play is, or what their part is. When it comes to "natural" behavior, many of us feel that same way.

So, how are we to act, what is appropriate behavior? I was reading a novel the other day in which the main character, a biologist, wondered, "We can study animals in their natural habitat to understand their natural behavior, but what is a human being's natural habitat?" Is New York or Tokyo or London anything like a "natural habitat"? Or even the small towns of Norman Rockwell's artistic imagination? We know that animals in zoos exhibit radically different behavior than those in the wild. Are humans living in zoos now or is this our new frontier? What is the natural behavior, the one true pattern that we can model ourselves after? Religions go about addressing this issue with as much fervor as biologists. Holy leaders provide blueprints, sacred guidebooks to behavior, that promise to surpass natural behavior (condemned as "animal-like") and lift us up to the spiritual. Yet, still we wonder: How should I act?

Now, back to the original question: Do animals have ethics? This book does not make the case that animals have ethics, since no such case can scientifically be proven. Scientists such as behaviorist B. F. Skinner and sociobiologist E. O. Wilson—and most recently writer Robert Wright in *The Moral Animal*—make convincing cases that human notions of morality are merely evolved out of practical biological needs, the same way all life forms physically evolve. Incest is immoral not because of a divine condemnation, but because it produces weaker offspring who require greater time and effort to care for, who will grow up less able to contribute to the social group, and who will weaken the gene pool of the entire group, threatening everyone's existence. If human morality is merely evolutionary, it is deterministic (the result of physical forces only), and therefore there is no free will involved. No real "ethics" as we have come to define the word. Human morality may simply be something we develop for survival in the same way the skunk has a noxious spray or the porcupine has spiny needles. To point to our notions of morality as indications of natural superiority to other animals would be like the tortoise pointing to its shell as proof of its natural superiority because it's got one and we don't.

That doesn't diminish the value of human morality. Indeed, such an understanding may help us perfect it, understand how necessary it is, even help us better to fulfill the promise it holds for our future. Ironically, I ended up learning more about the basis of human morality from writing this book than I did from the six years spent writing my religion book. The Marriage and Courtship chapters were eye-openers in helping me comprehend why men and women behave toward each other the way they do.

And so it is with all the chapters in *Kinky Cats*. By examining some of the major human moral terms (abortion, birth control, homosexuality, rape, sex, etc.) and how they relate to actual animal behavior—as well as to the "natural" explanation for such behavior—we may better understand our own morality as well as have a greater understanding of the broad spectrum of what's natural that most people haven't yet begun to fathom. Perhaps we will widen those paths of what we consider "natural" and come to understand ourselves as humans a little better.

That we should desire to do so is only, well, natural.

1
ADULTERY

Field notes. . .

Adultery does not so much involve sex as it does betrayal. Humans have trouble defining adultery in polls because we don't all agree on a specific act that constitutes adultery. For example, is it adultery to share a passionate spit-swapping kiss with someone other than one's mate? Is masturbation adultery, since it involves sex with someone other than the spouse? Is sex with a prostitute really adultery, since it doesn't involve love, only body parts? Some couples who openly engage in extramarital sex via spouse swapping still condemn adultery, which they define as any sex act with another that occurs without the spouse's knowledge; so for them, it isn't the sex that makes the act adulterous, it's the secrecy.

For adultery to take place, there must first be a clear relationship with the potential to be damaged by the act. That's why this section doesn't include animals with harems or those who have promiscuous sex without forming any relationships. In fact, many of the following examples of adultery could just as easily be placed in the Marriage chapter under the section on lifelong monogamy (the "Till-Death-Us-Do-Part" Marriage); and most of the species that appear there could equally appear here, since wherever there are monogamous marriages there's usually adultery. When we use the term monogamy, we're talking about a general pattern of behavior embraced by a particular animal society. But, as we know, that doesn't mean the pattern is strictly adhered to. Most of the human societies at present preach monogamy as the ideal lifestyle, yet adultery rates in many countries are enormous. In the U.S., the adultery and divorce capital of the world, recent polls indicate that 90 percent of Americans think adultery is wrong, yet surveys show that the rate of adultery is between 70 and 78 percent. Therefore, when this chapter focuses on the adulterous activities of an animal, it is usually within the context of a monogamous relationship. Sure, they have lifelong monogamy—mostly. They stay together as a couple until one dies, often with a devotion that human divorce rates indicate we can't match. But on occasion either spouse may have a "quickie" on the side.

BABOONS

While some baboon societies consist of large groups of eighty or more, the **hamadryas baboon**, which inhabits the desert area of Somalia, lives in a small harem consisting of the prime male, a minor male, and two or three "wives" (see also MARRIAGE: BABOONS). At night, all the individual harems come together to sleep in a huge group at a traditional sleeping ground. But come morning, it's every baboon "family" off on its own foraging for food.

In some harem-based societies (see MARRIAGE: FUR SEALS and SEA LIONS), the males must constantly protect their harem from other males who are trying to steal their females. But the male hamadryas baboon has a code of ethics concerning other males: They will not attempt to steal any females from another male—as long as they recognize the male as one of the guys from the same sleeping ground. It is unknown whether this code is the result of a male camaraderie, as in a fraternity or street gang, or simply because it would cause too much strife at the sleeping grounds.

FINCHES

Although some birds only mate once a year during a specific mating season, others are more sexually persistent and mate throughout the year. The finch is of the latter group. Despite this more lively sexual life, they tend to form into couple-centered relationships like human marriages. However, like human marriages, there are temptations. Sometimes when a new female appears, the male leaves his mate and tries to forcibly have sex with the new female. But his attempt is useless unless she cooperates. If she doesn't cooperate, he may try to court her the old-fashioned way, just as he did his wife.

The male's new obsession is not taken in stride by his mate. She will tag along wherever he goes and go through her own courtship rituals to win him back, including the traditional bowing, dancing, feather ruffling, and sexual crouching that worked before. At the same time, she may physically attack the new female, who generally doesn't fight back.

"Did you ever have to make up your mind," sings the Lovin' Spoonful, "to take up with one and let the other one slide?" Same dilemma for this songbird. In the end, he must choose one over the other. If he returns to his mate, the two of them may team up and attack the new female, driving her away. If he prefers the new female and begins copulating and nest building with her, the new mistress will now start to fight back when the old mate attacks her. And the male will now join his new bride in physically driving off his now ex-spouse.

➾**FYI:** If the new member is a male that attracts the female's fancy, the same basic process occurs.

GULLS

The kittiwake gull prefers monogamous marriages for a very practical reason—studies indicate that monogamy produces a higher number of offspring, thus better insuring the survival of the species. Though mates separate for part of the year, a majority of them return to their previous mates for each new mating season. Still, there are temptations. While the happy couple is off fishing, an interloper may swoop into their nest. If the guest is a male and the male owner of the nest arrives first, the owner chases the guest away. But if the female is first to arrive, the male will attempt to seduce her. If the uninvited guest is a female and the male owner is first to return to the nest, the female will attempt to seduce the male. On occasion this bold technique works and the marriage is shattered. However, the longer a couple has been together, the less likely they will fall for this temptation. (See also DIVORCE: KITTIWAKES.)

2
COURTSHIP

Field notes. . .

Humans are obsessed with courtship—which we call "dating." Magazines such as *Cosmopolitan* and *Playboy* are thinly disguised courtship manuals that instruct women and men on proper courting ritual procedures. One of our most enduring art forms, the romantic comedy, tells the story of boy meets girl, boy loses girl, boy gets girl. In fact, the romantic comedy is widely referred to in the movie industry as a "date flick," because couples who are courting will be attracted to such a story. Girls hope such a story will instruct their dates on being romantic; boys hope it will put the girls in the mood to fool around. Advertising promises that with the purchase of most products (deodorant, mouthwash, makeup, clothes, cars), we greatly increase our chances of finding a suitable mate. If humans, who consider themselves the most rational of beings, are obsessed with courtship, you can imagine how much more intense other animals are.

Courtship rituals vary, but they all exist to serve the same basic purpose: find a suitable mate for procreation. But the variations and nuances of the rituals reveal that this is not an easy matter. If mating were just a case of delivering sperm to an egg, pretty much any male or female would do. We could eliminate courtship altogether. But we know that different species have different degrees of selectivity about their mates. The more selective a species is, the more complex the ritual. Courtship is a competition—an Olympic event in which the winner gets to mate and the loser gets to watch. In fact, our own competitions, including the Olympics, are nothing more than variations on courtship ritual. Mostly it is the young focusing their mating energies into perfecting one form of activity that will impress others, especially potential mates. This instinct is present even among those who may already have a mate.

The reason individual animals are so selective is that they are not choosing a mate just for their own needs—though they think they are—they are choosing a mate that best serves the entire species. They are selecting a mate who will produce the offspring with the best chances of survival. This in turn promotes the survival of the entire species. The complexity of the courtship ritual among some animals is so demanding that only the strongest and most agile of males mate. Those who don't make the grade don't get to mate. They only get to watch the winners pass on their genetic code.

Within a species, the courtship ritual is designed to solve three problems: (1) it allows members of the same species to recognize each other; (2) it allows members to recognize members of the opposite sex; (3) it allows members to develop their sexual maturity at the same pace as those of the opposite sex. As you will see, it is not uncommon for some animals to be unable to tell males from females in their own species. Even more confusing, some animals don't even know their own sex. Courtship rituals allow them to play the roles of both female and male and realize which feels most comfortable. Once these questions of sexual identity are resolved, the next function is to have the courting couple develop sexually at the same pace so they can produce offspring. The courtship ritual actually triggers chemical changes within the bodies of animals to make this possible.

The courtship ritual puts the dating couple through an intense basic training meant to reveal strengths and shortcomings in a condensed period of time. The time period has to be condensed for most animals because they must make their mating choice while the female is in heat. Sometimes this period is only a day, sometimes a couple weeks. Elaborate courting rituals are especially necessary among animals with a higher level of aggression—like humans. (A species' level of aggression is caused by their physiological makeup, not their choices. Humans have religion and philosophy as a means of channeling and controlling their natural aggression. But some animals behave according to our notions of piousness because they lack the biological levels of aggression we have.) In a fight or flight situation, aggressive animals tend to attack first. Among many prehistoric dinosaurs, rape seems to have been the normal method of sex. The love bites that are so familiar among today's animals, from lizards to humans, were a bit more lethal back then. Even among vegetarian dinosaurs, skeletons have been found showing teeth marks that were caused, not by carnivorous predators, but by other vegetarians of their species. Some scientists theorize that the reason for this violent sex was that dinosaurs had not yet developed ritualized courtship, which allows animals to prove they will not be aggressive toward each other. Without that, the male may have decided it was best to strike first.

This aggression spills over to their intrapersonal relationships. Because many animals can't recognize members of the opposite sex, they first go into attack posture. The courtship ritual finally convinces them that there is nothing to fear. One of the main issues to be resolved during courtship is trust. Food is the symbol of such trust. The male has to trust that the female won't steal his food and the female has to trust that the male will provide food for her and her offspring while she guards them from predators. The courtship ritual sometimes is nothing more than a demonstration that each cares more about the other than they do about food (see TERNS below). This ritual is called "courtship feeding" (see also PROSTITUTION). Certain male animals, such as a **quail**, will offer food only after other methods of courting have failed. In other cases, food is offered as a distraction to avoid being eaten by the female (see SEX/DANGEROUS). Sometimes males will bring much more food than is necessary or even possible for the female to eat. The **woodchat strike**, for example, continually delivers food to his mate like a maniacal Dominos pizza delivery person. Back and forth he flies, rarely resting, rarely eating. Day and night delivering food. The **red-crested pochard** is another bird obsessed by his desire to feed his mate. For months he will dive ten feet underwater in order to uproot the plants she needs. He is so driven that she doesn't even get a chance to finish what he's brought her before he arrives with a whole new course. The reason for this behavior is less a bribe or show of ability than it is a means by which the males use up their sexual energy while waiting for the females to come into heat. Without this "release," the male might otherwise turn this sexual energy into aggressive behavior, such as rape.

Through courtship feeding, animals demonstrate the principal of the human concept of salvation: delayed gratification is more rewarding than instant gratification. Courtship rituals are a way for each animal partner to prove to the other that he or she shares this same "family value."

AKOUSHI

The akoushi is a South American rodent slightly larger than a guinea pig. When it comes to mating, the male has a significant challenge: The female is able to conceive only for a few hours. So he doesn't have time to orchestrate a big wooing campaign. He's like a door-to-door salesman with one foot wedged in a closing door; he's got to make his pitch before the door slams on that foot. Once he finds a mating candidate, he launches into his baby act—so effective among many different species, including humans. He cries out like a hungry child until the female's maternal instincts are stimulated. When she approaches him, he begins to shiver, thereby proving how helpless and non-aggressive he is. He may even pretend to run away from her in fear, which he does by turning his back and running in place. Apparently, this does the trick. The female accepts this "vulnerable" male as a mate.

When the baby act doesn't work, he has an alternative plan, not quite as benign as the first plan. He stands on his hind legs and shoots a jet stream of urine at the female until she's soaked. This marks her as his property and other males stay away from her. Because she only has a few hours in which to mate, she has no other choice now but to mate with the male who gave her the golden shower.

➾**FYI:** The **agouti** is a relative of the akoushi, but its courtship method is much more direct: The male chases the female until she collapses. Out of exhaustion she accepts him as her mate, though it is not rape. The human variation of this is nagging, to keep asking someone out until they finally say yes.

ALLIGATORS

Surprising to some, sex between alligators is a fairly gentle affair. The male may court the female for several days, clawing at her behind, then snuggling his head against her throat. Sometimes he dives beneath her in the water and acts like a Jacuzzi, emitting bubbles to tickle her cheeks. He also is able to pump small jets of water up through his back in fifteen-second durations. When finally aroused, the female begins to writhe about his body, over and under him. If both are excited now, they may snort and smack the water with their tails. Actual sex occurs in a somewhat side-by-side position, though facing opposite directions: His penis extends forward then shoots off to the side, allowing him to enter her at an angle. They remain in this position for about fifteen minutes with very little movement (because the sperm is rushing along the groove in his penis into her cloaca). Afterwards, the male may remain faithful to the female for the rest of the breeding season.

➾**FYI:** Despite his gentleness as a lover, the male alligator is a much more aggressive parent: If he happens upon the eggs his female partner will lay in two months, he will eat them.

BEETLES

Famed movie director Roman Polanski (*Chinatown*) was once convicted of giving drugs to an underage girl and having sex with her. The tradition of using drugs to get sex sounds like something out of our dark past, but it still goes on. The FDA has recently prohibited a European tranquilizer from being sold in the U.S. because it has become a popular modern "Spanish fly." As reprehensible as this behavior is among humans, there is a variation of this among beetles that is not only common, but necessary.

When the male and female beetle meet, the female is so frightened that she starts to flee. The male immediately spins around and presents his butt to her. No, it's not an insult. His butt actually contains an organ that produces—brace yourself—an aphrodisiac, which she greedily gobbles down. After a while, the male skitters around to her rear for sex. If she kicks him, he offers her more of his "treat." If she doesn't kick him, she offers him her treat. The substance that he feeds her actually matures her sexual organs, making it possible for her to have sex. This is similar to the "love apple" that was common in Elizabethan times: The woman would hold a slice of apple under her arm to absorb her odor and present it to her lover as a gift.

➠**FYI**: The male **scorpion fly** also manufactures his own food in the form of saliva. He feeds the female his saliva, which he has secreted onto a leaf where it has hardened into a tasty food. She devours it greedily while he sneaks up and mates with her. This is how he avoids being eaten by her.

BOWERBIRDS

In the movie *The Housesitter*, Steve Martin drives his blindfolded girlfriend to a secret place, whips off her blindfold, and gestures with a huge grin at the house he has designed and built for her (he's an architect). It is at this precise moment he asks her to marry him. She turns him down. But Steve is not alone in using his architectural skills in an effort to court a potential mate. The bowerbird (New Guinea's bird of paradise) also attempts to attract a mate by constructing elaborate cottages, not to be lived in, but merely for sexual conquest. Some of the buildings are merely huts built from a few branches. But others are multiroomed structures so elaborate that early explorers mistook them for natives' homes.

Some bowerbirds forgo the house in favor of stagelike structure. The **Sanford's golden-crested bowerbird** builds a circle, which he decorates around the rim with colored snails' shells and the colored wings of beetles. He also hangs a curtain of bamboo sticks and fern leaves. He then dances around the stage, singing and posing until a female decides to accept the invitation. Once she does, they go behind the curtain and copulate.

The **golden bowerbird** is known for its magnificent Maypoles. Though he is only about nine inches long, his Maypole may, after several years' additions, reach nine feet in height. The process starts when he makes a six-foot clearing in the forest and jams a tree branch into the ground, to which he attaches twigs until it resembles a small tree. When a female joins him, they dance around the "pole" until they reach an excited state. They then retire discreetly to a smaller house, which has been decorated with orchids, bright flowers, lacy mosses, as well as shells and berries. When the flowers wilt, he replaces them with fresh ones. It is here the couple consummates their relationship.

Most impressive of all, however, is the "avenue" builder, such as the **regent bowerbird**. After clearing a space four feet in diameter using sticks, he drives sticks into the ground in two parallel lines like a hallway. Then he bends the tops together and ties them with vine and plant fiber, forming a long archway just wide enough for him to walk through without scraping his wings. But he is not just an architect and builder, he is also an interior decorator. The floor is covered with bleached bones, flowers, and colored pebbles. Finally, using a twig as brush and berry juice for color, he paints the walls.

However, despite the ingenuity of the male bowerbird, it is important to realize that not all males get the opportunity to show off their building skills. The female **bird of paradise** mates only with the strongest, largest, and most handsome of the male birds. This status is bestowed by conquering the best territories, thereby giving them the distinction of being permitted to build their mating houses. Other males who are not landowners, do not build houses nor even attempt to mate. This method of selection seems designed to improve the strength of the offspring to the benefit of the society of all bowerbirds, which is why less dominant males agree not to mate. One wonders how humans would react under such a strict moral code in which a majority of males agree not to mate so that the stronger ones could in order to produce more fit offspring. (See also VIOLENCE: TURKEYS.)

After mating, the male loses interest in creative behavior and does not join in raising the offspring. The glorious houses, which are used for courtship only, are abandoned and the females go off alone to build small plain nests and rear their young. (For an interesting variation on this courtship through nest building, see SIAMESE FIGHTING FISH below.)

➾**FYI:** Some researchers believe the behavior of the bowerbird is an indication that they may be evolving toward monogamy. They don't have the gaudy plumage of some birds that attract predators and results in them being banned from the nest (see MARRIAGE). There is no reason why the males shouldn't stay and raise the young. In fact, the elaborate building may be an artistic representation of actually having bright plumage. Perhaps they evolved away from possessing the plumage and replaced it with their building prowess. What's interesting in this case is the development of art—something that represents something else. This is not unlike humans telling stories—through TV, novels, or films—that are entirely fictitious, yet they move people to emotion and thought, even to change their perspectives on life.

CUCKOOS

Despite attempts to purge social rituals that reflect sexism, human society still encourages the ceremonies founded in prehistoric need. When a boy and girl go out on a "date," the boy is generally expected to pay. In fact, the boy is still expected to initiate the whole date in the first place. And a box of chocolates and some flowers wouldn't hurt, buddy. Perhaps those rituals were necessary at one time, but are they still? Or are they now just a matter of habit?

When the male **European cuckoo** goes courting, he offers the traditional mating gift of an insect. But why? Offering such a gift generally demonstrates one's ability to provide food or shelter. But this bird is polygamous; he won't be hanging around long enough to feed anybody or build any shelters. Besides, the female doesn't need any shelters since she lays her eggs in other birds' nests (see PARENTING: CUCKOO). So, the whole mating gift idea is merely a tradition from days long ago when the European cuckoo may have had to prove himself as a decent insect-winner.

DUCKS

Duck courtship is reminiscent of a high school football game. The males perform an elaborate ritual while the females gather around to watch and swoon—and select who they want to go steady with. The ritual is not for the weak; it is physically demanding, involving various aerobiclike movements as well as strenuous swimming. The routine and choreography is very precise, and any male omitting steps or making any mistakes is not going to the prom. When the males are performing especially well, some of the females may be overcome with team spirit and swim among the males nodding their heads up and down in appreciation, their version of "Go, team, go!"

Eventually, of course, team spirit gives way to the dating urge, the females begin to focus their attention on the individual males they are interested in, the cute one with the tight tail feathers. When two seem to like each other, they pair up and swim off together. Because the courtship ritual takes place in December, neither duck is sexually developed enough to actually mate yet. They are just going steady; if they find they don't get along, he takes back his team jacket and they both return to the ritual courtship grounds—he to perform once more for the Gipper, and she to swoon over another. Those who don't break up, mate in the spring. In general, ducks remain faithful to each other for life.

ELEPHANTS

The courtship of elephants is the stuff of romance novels. The female is coy and coquettish, running away from her chosen mate, but not so far or fast that he can't catch her. And unlike the quick slam-bam courtship of most animals, the elephants enjoy an extended period of flirting and heavy petting. They kiss, which usually includes their variations of French kissing: sticking their trunks into each other's mouths. Their trunks become major courtship devices: They will braid their trunks together like humans holding hands; and they use them to stroke each other, including mutual masturbation. But most important is that, during this period, they are inseparable in their daily activities and openly display affection for each other. However, despite their increasing sexual play, they do not suddenly leap on each other with lust. They wait until the female is ready to conceive; after all, she only has sex once every five years.

Once they have sex, however, the romance is over. Their sexual encounter can last for several days, but then each leaves the other. The female (cow) returns to the company of other females and youngsters; the male (bull) returns to the company of other males. (See also SEX/DURATION & FREQUENCY: ELEPHANTS.)

FIDDLER CRABS

You've seen the photos of the fiddler crab with its enormous single claw curved in front of it like the bumper on a scuttling VW bug. Well, only the males have those giant claws and they're more of a nuisance than anything else since they actually make it more difficult for the male to feed himself. Still, they are necessary for attracting females, so the male lugs it around and cleans it constantly (which makes him look like he's playing a fiddle). The male also attracts females through his changing colors. The female is always a drab brownish gray, the better to be camouflaged in the sand. The male is this same color when he first awakens in the morning or if he's frightened. Otherwise, he is as colorful as a seventies disco dancer, with the hues of his body constantly changing. The big claw and colorful display are all for the purpose of attracting females. When he is successful and a female shows interest, he begins a frenzied dance, pulling her along with him in an effort to arouse her sexually—just like human dancing. To help excite her even more, he begins stroking her legs; if interested, she returns the favor. Once the female is sufficiently aroused, the male leads her back to his love pad, which is a hole in the sand. He scoots inside, she follows, and he quickly plugs up the hole with a swatch of mud to insure a little privacy.

GERBILS

Humans go on the *Love Connection* with a long list of all the qualities they are looking for in a perfect mate, only to be thwarted again and again in their pursuit. Gerbils also have a list, though it's short and easily filled: big and smelly. The bigger and smellier the male is, the more in demand he is as a mate. Experiments by psychologists Mertice Clark and

Bennett Galef from McMaster University in Ontario, indicate that a male's virility is determined by his location in the womb during gestation. They found that males who gestated between two males, were born with higher blood levels of the male sex hormone testosterone. This is because the male's testes secrete testosterone during the latter stages of gestation. Apparently this testosterone leaks into the adjoining amniotic sacs of the siblings, so that a male gestating between two other males gets a major dose. Once born, this middle male grows faster and bigger than either his siblings or other males who developed between sisters. Also, the middle male's scent gland for marking territory is bigger. When given a choice between these testosterone-charged males and those who gestated in a less enviable position, female gerbils consistently choose the larger and smellier male. This method seems to make sense, since these king-sized gerbils sired larger litters than their rivals.

➾**FYI:** Female gerbils gestated between two sisters also matured faster and gave birth to larger litters.

KISSING GOURAMIS

The sight of these fish lip-locked in a puckered kiss is enough to elicit a smile from even avowed nonromantics. But don't be too quick to go all squishy inside; their kiss is as much an expression of aggression as it is romance. Once they stick their lips together, they suck until they are stuck that way, then they begin a violent thrashing of tugging and pushing. This can continue for hours, with the two of them breaking away between rounds, only to return for more of the same. If neither fish is able to defeat the other, this is a sign that they are compatible. They then separate, mate, and return to their kiss again, this time more gently.

KOB

A male friend of mine who owned a $200,000 Ferrari told me he used to come back to his parked car and find notes from women who had left their phone numbers for him to call. These are women who had never even seen him, only his car. Apparently a pricey penthouse, condo, or mansion can induce the same aphrodisiac effect on some females. The same holds true among other animals, including the kob, a type of antelope that lives in Central Africa. The male kobs who dominate the best pieces of land get all the females; those who don't, don't have sex.

The kob designate a certain section of land (about the size of two football fields) as the mating grounds. Here, and *only* here, sex will take place; males will not attempt sex anywhere else, and females will not consent to sex anywhere else. A herd of about one thousand kobs will gather in this area, which is divided into roughly fifteen plots of land, each nearly fifty feet in diameter. The land itself is closely cropped green grass, as fine as any manicured lawn in Beverly Hills. But all plots are not of equal value; the most valuable sections are in the middle, with the outer areas less desirable. During the mating season, the males fight each other every day over the prime sex plots, each trying to move closer to the precious center locations. Those males without any property constantly battle those around the perimeter plots, looking for their way in.

When the females come into estrus, they walk into the mating area looking for the buck with the best genes. Naturally, the ones who have battled their way to the best locations look to them to have the best genes. Still, it's a very formal process, with nothing guaranteed. The doe may walk onto the property of a potential mate and begin grazing. The male then comes over and places his forefeet between her hind legs. If she doesn't move, he mounts her. However, she may not be interested in him and walk off to another prime plot of land. The male doesn't try to stop her, nor does he fight with the lucky male she chooses. If he holds the land, they will come. A typical day for a male in a good breeding spot will consist of about twenty fights with male challengers and twenty copulations with willing females. (For a similar use of prime real estate as means of establishing who has sex, see SEX/DURATION & FREQUENCY: SAGE GROUSE and MARRIAGE: GNUS.)

LIZARDS

"If you want to catch a man, let him think he's the boss." Familiar womanly advice throughout history, this courtship through deception attempts to address what is often called the "fragile male ego." Humans argue whether such means tarnish any ends that result, but animals don't seem to anguish over any means that achieve necessary ends. Hence, if the female lizard wishes to mate, she must make the male lizard feel like the master of his domain, otherwise no mating can take place.

First, the male establishes his territory, which he defends against any intruding males. If he encounters an interloper, he will do rapid pushups. Usually this is enough to drive the foreign male away. If a battle does ensue, the home team almost always wins. When a female enters his territory, he approaches her with head high so she can see the bright colors on his throat. She may run off at this point. But if she stays, she will acknowledge his dominance by stamping the ground with her forefeet; this is the same movement that a defeated male makes following a losing battle. She then runs off and he gives chase. However, she doesn't try to outrun him, just stay enough ahead so he can lick at her tail. In fact, if she slows down too much, he may nudge her along. Unlike some species in which the male is generally larger and stronger, the female lizard doesn't have to endure this treatment. She has the size and strength to fight him off; but she chooses to acquiesce to his dominance, since she knows he will not mate with a female who is too aggressive.

Next, he grips her neck in his jaw, and jumps onto her back. She raises her tail and he inserts one of his two penises into her cloaca. Several thrusts later his tail quivers and he is done. Because his penis is barbed, should the female start to run off too soon, she will drag him behind her, his penis still attached.

MAYFLIES

Life is short, we are often advised, gather ye rosebuds while ye may. Indeed, for the mayfly this ain't no metaphor. After living several years as a grub, it metamorphoses into a flying creature, like a butterfly. One problem: It now has only twenty-four hours to live. During most of that time it performs an aerial courtship along with the millions of other mayflies that are all around. The flight pattern is very precise: The mayfly flies up eighteen inches, floats down eighteen inches, then flies back up again. This is repeated until dusk, at which time the male and female come together in sexual union. Thus attached, they continue flying this same eighteen-inch bobbing pattern. After the sun sets, they break apart and the female zooms low over the water, pushing out eggs from the twin tubes at her hind quarters. Shortly after that, the mayflies all die.

MOSQUITOES

At the beginning of the film version of *West Side Story*, the Sharks and the Jets show their individual street gang identities by each singing and doing an elaborate dance. They look cool and sexy, as their women attest by running into their arms afterward. Same deal with mosquitoes. In some species the males congregate in large swarms of millions high in the sky, but only at dusk or dawn. Then they take off as a gang on the prowl for females. But they don't just cruise—they serenade. They hum in unison as they fly, the largest choir in the world, looking for love. When females hear this insect Gregorian chant, they come zipping out of the grass and head straight for the love gang. As the females near, individual males begin to veer away from the group, looking for their own date. Sexual contact is made in midair, but they usually fly off, still joined together, to someplace more private to finish their copulation.

MOTHS

Humans know the importance of scent in the mating game. We used to cultivate the gritty, more sensual odors as part of our courtship. In Elizabethan times women used to send their male suitors a "love apple," which was the fruit pressed under their arms to absorb their sweat. Today, Americans prefer to sanitize themselves to the point that very little personal odor might be their signature smell; instead they douse themselves with scented deodorants, perfumes, lotions, colognes, etc. Obviously these concoctions are designed to merely attract one person to the other, after which the force of the individual's personality will take over. But what happens when the scent becomes the end in itself?

The *Attacus atlas* moth is one of the most beautiful specimens in the world, with bright colorful wings that span ten inches. The female sits on a branch and sends out her own fragrance to attract males. And it does just that. They mate and all is well. However, experiments have shown that the male moth desires only the scent and is not attracted by any other characteristic of the female. It's like a powerful aphrodisiac. Zoologists have doused tissue with extract of the female's fragrance and watched as males swoop down to try to mate with the tissue. And all the while a gorgeous female sits nearby, but in a glass case so the male cannot smell her. She was ignored.

➾**FYI:** Similar studies with mice indicate that selection of mates is done primarily based on smell; but what the mice are smelling, and are attracted to, is a DNA structure that is far different than their own. A couple with similar DNA structures that mate produce smaller and weaker offspring with less of a chance for survival—hence the general taboo against incest. Therefore, a mouse smells those members of the opposite sex that also have DNA that is opposite and pursues them.

PEACOCKS

We have seen variously throughout this book, that mating often times requires a gift from the male to the female (see PROSTITUTION). Whether this is prostitution or just a show of affection, sex will often not take place without such a gift. Usually the gift is food, though among some birds and insects the gift is more symbolic than practical; a stone may be substituted, or the food is never eaten, merely passed back and forth to prove that neither would eat the other's food. In some cases—certain spiders—the gift is a sham, all packaging that once unwrapped is empty. However, by the time this is discovered, the male has mated and run off.

Male peacocks can be just as deceitful. In general, the male keeps a harem of four or five hens. When he wants to mate, he doesn't chase one down; he merely wanders over to the courtship area of his property and spreads his colorful tail. Though this display may seem akin to a rock singer grabbing his crotch and saying, "Let's do it, baby," what the male is really signaling is that there is food present. When the female sees his signal, she comes charging over for food. Despite the fact that he's lied, the male still tries to mate with her. Sometimes it works.

PENGUINS

If you've ever been in a crowded bar or dance club with wall-to-wall bodies trying to figure out how you'll ever manage to meet somebody—let alone a perfect somebody—you know what the **adelie penguin** goes though. Every year a hundred thousand or so of them gather on the coast of Antarctica to breed. The problem is that their eyes are designed for seeing when swimming underwater, not when hopping around on land. So, when they all get thrown together, it's the equivalent of looking for that special someone in the smokiest disco club around. How do the males meet the females? Same way guys in bars toss out some lame line or offer to buy a drink. The penguin equivalent to a line like, "What's your sign?" is a loud braying, croaking sound. This is necessary because of the poor eyesight thing; to penguins other penguins all look alike, even the males and females. Once a male spots what looks like an available female, he rolls a stone along the ground with his bill until he's left it at her feet (the equivalent of buying her a beer). The recipient of the gift might then begin screaming angrily at the gift giver to tell him that they are both males and he'd best move along. The recipient might also pummel the gift giver with her wings, rejecting his offer. Or she might start singing along with him. If she does, she will also start dancing around him, and they will sing together awhile before mating. This pebble will become the first "cornerstone" of their nest together.

SIAMESE FIGHTING FISH

These are the fish that keep appearing in movies like *From Russia with Love* and *Rumble Fish* because of their nasty habit of attacking each other with such ferocity. They provide the perfect metaphor for human aggression. But even more interesting than their fighting ability is their unique nest-building ability. They, and their relative, the **paradise fish**, possess a breathing apparatus in their heads that allows them to extract oxygen from the air. This is accomplished by swimming to the surface and sucking in air through their mouths. But during mating season, they don't use this to breathe; instead they coat the air bubble with saliva and let it float to the pond's surface. Hundreds of these spit-coated bubbles mesh together to form a frothy nest, which the male then floats to the land. This nest is his Porsche, mansion, and thick wad of cash all rolled into one; it is the basis by which a female will chose him as her mate. He who has the coolest bubblenest gets the girl. After he's squeezed her eggs out and fertilized them, they are enclosed in the bubblenest. But his work is not yet done. Like a Sunday afternoon fix-it dad puttering around the house with his tool belt, the male fish must keep manufacturing bubbles to replace those that continually burst.

TERMITES

Termites live in elaborate underground communities, and though they have eyes, they are virtually blind—except for the day when they will mate. Yet, for this day to take place, conditions must be perfect, requiring more conditions than Donald Trump's prenuptial contract. First, when breeding time comes around, the termites in all the nearby hills fly up into the air at the same time. This condition makes sure that they won't interbreed. Second, the lighting must be just right. This isn't a matter of tossing a red scarf over the lamp for the right mood; it's a matter of practicality: It must be light enough to find each other and dark enough that they aren't attacked by predators. Third, it can't rain while they're all zipping around in their mass mating flight, but it has to rain soon afterwards so the ground is soft enough for them to burrow in for shelter. When everything is finally perfect, the worker termites open the special exit at the top of the mound. Soldier termites skitter out to secure the area, to make sure no predators are lurking. If all is clear, the soldiers release a scent that tells the others the coast is clear. Then several hundred thousand termites from all the neighboring colonies fly upward like churning black clouds and the mating begins.

TERNS

Ah, the age-old question for those with wealth: How can I be sure my prospective mate loves me for who I am and not just for my money? The terns have developed a courtship strategy that addresses this question. Though terns are related to gannets (see VIOLENCE: GANNETS), unlike the gannet, they are a sociable bunch whose rituals involve gentleness and are therefore more symbolic than the directness of the daily beatings the male gannet bestows upon his mate.

The male tern who is in a mating mood first catches a particularly fine fish, then he marches among the eligible females, displaying his fish. When he sees a female he likes, he offers her his catch. The female must now proceed with caution; if she takes the fish, they are in essence wed, as if she'd taken his engagement ring. For this reason, despite the male's fine fish, the females may still decide to reject him. Not a quitter, the tern may then just eat the damn fish himself and fly off to catch another, better fish to woo the females—like trading in a Honda for a Porsche. Generally, however, one of the females will accept his offer. However, she doesn't eat the fish; she holds one end in her mouth and he holds the other end in his mouth. They may stand this way for over an hour. This demonstrates that neither is greedy, that they can both control their selfish appetites for the common good of their relationship. This very ideal is the basis among human religions not only for marriage, but for marriage as a symbol of people's covenant with their gods: They promise to curb their desires brought on by being in material bodies in order to nourish their spiritual selves.

➾**FYI 1:** The **Bohemian waxwing** has a similar ritual, except it uses much smaller objects such as a berry, a fly, the pupa of an ant. These things are too small for both waxwings to hold at the same time, so they take turns holding it in their mouths, then pass it back and forth to each other. The important thing is that neither actually eats it; instead they prove that their devotion to the other is stronger than to their desires. In fact, neither will eat the food they've passed back and forth. They prefer to throw it away.

➾**FYI 2:** The male **blue-faced booby** goes a step further than either the tern or the waxwing. Instead of using food, he uses a stone. One reason for using a stone is that the ceremony takes so long, anything organic might have spoiled or be shredded by the time the whole thing is over. Like the Japanese tea ceremony, this ritual is not about practical ends—drinking tea or eating a fish—it's about patience and self-control. As with the tern and his fish, the booby struts around with his precious "gem" the size of a chicken egg, displaying it to all the eligible females. He will then drop it at a particular female's feet—and wait. If she ignores him, he'll move on down the line, dropping it at the feet of whomever else he's interested in. An equally interested female will then snatch the stone up, shuffle away a few steps, and drop it. The male grabs the stone again, offers it to her again, and she again snatches it up, walks, and drops it. This goes on for a couple hours until she finally accepts the damn stone once and for all. Now they are mated for life.

TURTLES

Mating for turtles is logistically difficult because the male must mount the female from the rear; that means he must somehow balance his flat underbelly atop her rounded shell. That can prove trickier than anything in the *Kama Sutra*. Fortunately he has a long penis, one fully dedicated to its single task of delivering sperm (it carries no urine). Still, for everything to work properly, the female must cooperate (with some exceptions—see RAPE: TORTOISES) by sticking her rear out from the shell as far as possible. To encourage this cooperation, the male turtle engages in some gentle courtship behavior. Usually, he walks directly toward her like a guy in a bar about to ask a strange woman to dance. If she's interested, the female turtle holds her ground and they stand there facing each other, bobbing and nodding their heads, sometimes for hours. As he becomes more confident, he may nibble at her toes or he may reach over and stroke her cheek. These gestures of affection may convince her to engage in sex.

3
DIVORCE

Field notes. . .

Okay, technically animals can't divorce since they can't legally marry. But nature isn't inhibited by such frivolous technicalities. Divorce by any other name is just as acrimonious and permanent. With the divorce rate in the Unites States at the 50 percent rate and rising, there is much head scratching about the whys and wherefores of divorce. Some biologists suggest that, whatever the apparent reasons humans give or convince themselves of concerning their own divorces, the real reason is the same among humans as it is among other animals: bad vibes. Somehow one of the couple senses the other will not make a good parent. The reason the couple got together in the first place was to fulfill their biological imperative of mating and producing offspring with the best possible chance of surviving and carrying on their DNA codes. Strip away arguments about toothpaste caps left off, messy bathrooms, too much TV watching, and you end up with the same basic reasoning. Remember, even if the couple is childless and intends to remain that way, the basic biological programming and urges are still operating, even if below the conscious surface. We are all biological organisms designed to perpetuate the species. Divorce occurs, even among supposedly monogamous species, when the behavior of one mate threatens this imperative.

Marriages rely on a sympathy bond between the two animals. This bond is established in various ways, often through an elaborate courtship ritual (see COURTSHIP). The ritual is important as a demonstration to each partner that it will not be harmed by the other; it also sometimes allows each to discover what its own sex is, as well as that of the other. Some animals establish this bond through smell; they recognize the other, not by appearance, but by smell. Some parents cover their offspring with a scent so they can distinguish it from anothers' offspring. If their offspring loses that scent, the mother may mistakenly kill it when she sees it. Other animals, especially birds, recognize their mates by sight, and the sympathy bond between the couple is based on that appearance imprint. Physical disfigurement can cause divorce because it affects the recognition factor of an animal. After courtship animals come to recognize their mate and the sight of the mate triggers a repression of their aggressive tendencies. But if one alters the mate's appearance, sometimes even slightly, it's as if that mate is no longer recognizable, and therefore the aggression level increases. To the animal booting out the disfigured mate, it is a stranger that's leaving, not their former mate. Experiments done with **herring gulls** demonstrate the importance of this recognition factor. These gulls recognize their own species as well as the opposite sex by the color of the ring around their eyes. Scientists who slightly altered the shade of the ring were successful in destroying longtime marriages among gulls because one mate no longer recognized the other. Even darkening a color is a deformity.

Divorce is a practical step for some species of animals. If they do not get along, they do not have offspring. Somc animals with very brief, aggressive courtships may have higher divorce rates because they didn't get to know each other enough to find out if they would be compatible. But haven't our mothers been warning us about that for years?

GULLS

If ever there was a metaphor among the animal kingdom for the "masks" between human man-and-wife relationships, it is the behavior of the **black-headed gull**. During autumn and winter these gulls look very much like all the other gulls, but during mating season they grow ninjalike black masks that are frightening not only to other types of gulls but to their own species. Because they are so frightened by the masks, they keep at a distance from each other, even when it is time to mate. To overcome this formidable obstacle, when the male calls and the female comes, they don't look at each other. They face the same direction and copulate without ever looking at the other's face. Since gulls are monogamous, they return to the same partner every year during breeding season, but still avoid looking at the other's face. If they do look directly at each other, they end up fighting or "divorcing."

"Divorce among gulls gives us some interesting insights to divorce among Homo sapiens," claims biologist Dr. William Jordan. "It shows, for example, that there is such a thing as fundamental incompatibility." With gulls this can occur when they disagree on who gets to incubate the eggs. They end up arguing, expressing themselves in the form of a loud choking sound. Because they spend so much time squabbling, the welfare of the offspring is at greater risk. The amount of time spent arguing takes away from time spent gathering food. Or, if the arguing becomes particularly heated, eggs can be crushed or the chicks injured.

Once divorced, the gulls arrive the following year at the mating grounds with new mates. The new mates invariably do not want to incubate as badly as the previous spouse.

➾**FYI:** As mentioned above, among many species of animals physical disfigurement can be grounds for divorce. Though the **kittiwakes** generally form a lifelong monogamous marriage, these marriages can be dissolved through an extramarital affair (see ADULTERY: GULLS) or if one of the partners suffers some physical deformity. One study documented the relationship of a couple that had been together for five years. One day the female returned from fishing with tattered feathers suffered in a fight with another bird. She greeted her mate with the usual rubbing of the beaks; but this time, instead of the greeting bringing them closer, both became agitated and they began pecking at each other. The relationship was over. She took off and within two days he had another mate in the nest.

But, like many humans, the longer the kittiwake couple has been together, the more likely they will stay together. Also, there are limits to what constitutes grounds for divorce. Disfigurement is a cause for divorce, but senility, sexual dysfunction, or sickness are apparently not. In these circumstances, the couple remains together, probably because the afflicted is still recognizable as the original mate and the animals can maintain their sympathy bond.

Divorce also has risks. Parents learn their parenting skills through practice and teamwork. The more frequently the kittiwake female divorces, the less capable she is as a mother, simply because she hasn't had the experience. Couples that return to the same mates are more experienced at raising their offspring, and their young therefore have a much better chance of survival.

➾**FYI:** Other animals, including **bullfinches**, **Bourke parrots**, and **violet-eared waxbills**, also divorce based on physical deformity. They are all monogamous birds that will endure sickness, senility and impotence—but not a bad-hair day. The **little ringed plove** is another bird that practices seasonal monogamy, returning each mating season to the same mate. Observers of a particular group noted that one couple had remained married for three seasons, but in the fourth year the male returned missing a leg. Although he was in no way impaired either sexually or in his ability to gather food or even move around, his wife immediately dumped him.

4
DOMESTIC ABUSE

Field notes. . .

Throughout this book you will find many examples of what humans would refer to as domestic abuse: one member of a sexually involved couple deliberately being violent with the other. Forms of abuse include everything from beating to murdering and eating one's mate. Usually abuse is directly related to size. In species in which the female is larger, she is the abuser; when the male is larger, he is the violent one. But this isn't always true. Sometimes the stronger male is beaten or killed by the less powerful female (see SEX/DANGEROUS: TIGERS).

If we followed the human definition of abuse, this chapter would be the largest in the book. However, many of these instances of abuse are not strictly reflections of the status of the relationship. In some cases, violence occurs during sex because it is necessary in order to physically copulate; the violence triggers a chemical reaction in the female that makes sex possible. In other cases, violence occurs because the animals are incapable of distinguishing between a mate and food. This makes bonding impossible, which results in every confrontation being a fight for life. Abuse also occurs when males struggle to maintain their harems or try to keep females from leaving their personal breeding grounds.

For the purposes of this chapter, I've only included only a few examples of domestic violence that are directly related to human notions of violence within a consensual mating relationship.

GANNETS

The goose-sized gannet is an extremely aggressive bird who must constantly fight for everything in order to survive. The problem is that the male doesn't have the ability to turn down its level of aggression and regularly turns it toward its mate in one of the most enduring examples of domestic abuse around.

Every winter the male gannet returns to its cold breeding grounds in the Gulf of St. Lawrence, and on the coasts of Newfoundland, Iceland, the British Isles, Brittany, and Norway. Immediately he searches out the nest that he and his mate had lived in the year before. Once he finds it, he may have to battle those birds that have decided to take it over. If he fails to win it back he moves on to another nest and claims that one for himself. Now he must defend his new home against the former owner and any new birds who take a fancy to it. Fighting between gannets is particularly brutal. The male gannet grabs his rival's head in his beak and yanks and shakes until both plummet off the side of the cliff. Each will pull out of this drop before they smash into the rocks or water below, but by the time they return to the nest, there may be yet another bird claiming it for his own. Fighting may continue now for a couple hours before one wins the nest. Victory may be sweet, but it is also short, for no sooner has a gannet captured a nest than other arriving males may challenge him to a duel for it. And so it goes for days. Once a male gannet has claimed his nest, he can't leave it to fish for food or find building material to fix up the now tattered nest. He has to wait and guard the place until the females arrive.

Males and females live apart for most of the year, coming together only during breeding season. Yet, they live for about twenty years, and during this time they remain monogamous. Still, though it's not a portrait of what humans would consider wedded bliss, when the females approach the thousands of nests containing thousands of squalling males, they immediately know who their mate is from the previous year and fly directly to him. The male doesn't exactly reward her loyalty. His aggression level is still so high from all the fighting that he welcomes her with a beating. He smacks her around with his wings and pecks at her with his beak. She is just as large as he is and could easily defend herself; when the male is away she guards the nest and handily drives away usurping males. However, when it comes to her mate, she chooses not to fight back. In fact, she makes a point of turning her beak away from him so he doesn't see it as a possible weapon. After about twenty minutes, the male relaxes and his blows become playful caresses.

You'd think that after all the turmoil, the couple could now enjoy some domestic tranquillity. Not so. The male now flies off to hunt for fish. When he returns, the violence starts all over again. He beats her daily for most of their lives together. (Some researchers believe that what we describe as beatings are also a form of sexual foreplay, for they will sometimes engage in sex immediately after a beating, without the usual caressing.) When both are old, the beatings stop. Now when they return to their nest, they simply mimic the violence without actually touching.

➾**FYI**: Like the gannet, **cranes** are also monogamous and mate for life (which can be as long as fifty years), but, unlike the crane, they stay together the whole year around—and the male does not engage in violence toward his mate. Though they also have the instinct or urge to attack each other—just as in human marriages—they ritualize this attitude. Instead of beating his mate, the male crane does a dance. This dance is not just performed for mating, but occurs throughout the year.

ORANGE CHROMIDE FISH

Humans often express the need to exert themselves physically in order to vent their aggressions and hostility. "If I didn't have tennis," some have said, "I might kill somebody." The same attitude can be found among the male orange chromide fish of southern India, who maintains his domestic tranquillity only as long as he has another fish he can bully around, a kind of whipping boy. The male will on occasion beat up another fish in the presence of his mate. As long as he has this whipping boy around, his marriage is stable. But if he doesn't vent himself in these periodic brawls, he will turn on his mate and attack her. Once he does this, the marriage is over and she swims off—and any young offspring they have will be abandoned and left to die. If the couple are in a small aquarium, however, he will attack her until he's killed her.

➾**FYI 1:** An interesting variation of this aggressive backlash occurs when the male is himself repeatedly defeated by other males in border clashes. This male does not lash out; rather, he busies himself like some finned Felix Unger from the *Odd Couple* in obsessive domestic chores. He digs endless spawning holes, not only more than he will ever need but also much deeper. If these border wars leave him only a small plot of land, he will first fill his holes before digging new ones. One deadly side effect: In his maniacal compulsion, he sometimes fills in holes occupied by his own spawn, killing them.

➾**FYI 2:** The **Asiatic water buffalo** is similar to the orange chromide fish in that unless he has one or two weaker males around to bully, he becomes sexually impotent. It's not enough that he has a whole harem of sexy females; he needs to demonstrate his power and position in an aggressive way in order to perform sexually. Don't feel too sorry for the bulls that he bullies; when the head honcho dies, the next-ranking male takes over the harem. The **brindled gnu** also must continually engage in mock combat with his neighbors in order to vent his aggression; otherwise he might turn it on the females, who would run away, refusing to mate. Thus, battle is a necessary component of mating; it keeps the males from attempting rape, which would result in no mating.

TASMANIAN DEVILS

These marsupials get the first part of their name because they come from the island of Tasmania just south of Australia. They get the second part of their name because they are aggressive little devils. Only about two feet long, they look like miniature brown bears. But don't let their size fool you—they will attack and consume anything within their path, living or dead, that isn't able to defend itself from their onslaught. This violent personality trait overlaps into their courtship and marriage behavior.

The usual behavior of the male is to live the hermit's life, securing a large amount of territory and then spending his days and nights driving off any others of his species, male or female. Then, when mating season rolls around, the male suddenly gets a hankering for female companionship. As befits his loner ways, he doesn't go courting, he goes shopping. He scours the countryside for a female; when he finds one, he snarls and bites at her until he's herded her back onto his own property and into his cave. There he holds her prisoner, not even permitting her to get near the exit. This goes on for two weeks. During that time he does not have sex with her because she is still not sexually mature enough; apparently the emotional excitement of the capture and imprisonment actually stimulates her sexual development.

After they finally mate, the male's rule ends. The smaller, weaker female begins a campaign of merciless bullying of the male, which only gets worse when she becomes pregnant. When the offspring are born, both male and female are devoted to their marital and parental duties. They build a nest for the young by digging a hole in their cave and lining it with hay. During the next five months, the mother nurses her young as well as goes hunting. The father cannot do the hunting because he must stay to defend his family against intruders who would otherwise sneak into the cave and eat the young. This basic family structure continues for nine months, until the offspring are ready to leave (though they will not be sexually mature until they are two years old). Once the youngsters have left, there doesn't seem to be any reason for the parents to stay together. So they don't. They separate and live alone for the next two and half months, when mating season begins again.

5
HOMOSEXUALITY

Field notes. . .

Among humans, homosexuality has long been considered a moral issue. Homosexuals have been and continue to be persecuted, sometimes to the point of being beaten or killed. Those who oppose homosexuality do so on the grounds that it is "unnatural" because it doesn't result in the production of an offspring—therefore such activity goes against God's design. However, in the wild many animals engage in homosexual activity as part of their natural sex lives. Captivity encourages even more homosexual activity than would otherwise take place if members of the opposite sex were available. Male primates in zoos who are deprived of the company of females will practice anal intercourse with each other, with the "feminine" male assuming the female's crouching position. Not unlike what occurs among humans in prisons.

Studies in recent years have found evidence of a genetic link to homosexuality among humans, which indicates that it is not so much a matter of choice as it is biology.

BABOONS

Young male baboons engage in homosexual behavior, but this stops once they are old enough to find willing females. This is also true of human children, who engage in homosexual experimentation based more on availability than desire for the same sex.

➾**FYI:** In general, when humans greet they shake hands or kiss politely on the cheeks, or hug. Each reveals a degree of formality or intimacy. When male baboons in the same troop greet each other, they give a friendly tug on each other's penes (plural of penis).

BEDBUGS

Bedbugs use their daggerlike penises to stab through the backs of the females in order to ejaculate their sperm (see SEX/DANGEROUS and SEX/KINKY: BEDBUGS for fuller descriptions). However, there is evidence that homosexuality is widely practiced among male bedbugs. Scars on male bedbugs often reveal numerous penetrations by other males. Some bedbug species even have a special organ similar to the female's, with which they receive the sperm of other males. Oddly, some bedbugs may actually ejaculate their own sperm as well as that of their homosexual partners into a female. Why do they do it? Bedbugs cannot distinguish between sexes, therefore they will try to mate with anything that is the size and flat shape of a bedbug and is dark in color like a bedbug.

➾**FYI:** The inability to distinguish between sexes is not that uncommon. To recognize a potential sexual partner requires a highly complex nervous system that many small insects do not have. Therefore, they mate with anything that is the same general size, shape, and color as themselves. The male **housefly** has the same problem; it will land on other males, pennies, the heads of screws, anything that resembles a female. Like a salesperson, the more contacts one makes, the more potential for a sale.

DOLPHINS

Dolphins are very sexual creatures, enjoying frequent sexual intercourse (even when there is no drive to reproduce), as well as masturbation and homosexuality. The **bottlenose dolphin** has been observed to seem almost obsessively sexual. Even when only a few months old, males will begin to initiate sex with other dolphins—male or female. (See also MASTURBATION: DOLPHINS.)

EARTHWORMS

The earthworm may resemble a tiny penis, but it is in fact a hermaphrodite. It has both the male and the female genitalia, so when it meets another earthworm there's no fumbling around trying to determine the other's sex. They simply latch onto each other at the collar that circles their midsections. After some side-by-side foreplay, they ejaculate into each other's sperm pockets. Much later this stored sperm will be used to fertilize their own eggs. So, although technically this is a strictly homosexual encounter since it involves only the male organs, it is one that eventually results in pregnancy. (See also SEX: GENDER BENDER.)

GEESE

Some male geese form homosexual relationships, preferring each other's company to that of a female. However, while the two are courting, a female goose may come between the two males and become pregnant by one of them. Both males accept her and the three of them form a family to raise their offspring.

WHIPTAIL LIZARDS

Some varieties of the whiptail lizard (also called racerunners) are completely without males. Like the legendary tribe of Amazon women, these reptiles are all females. Through virgin birth (parthenogenesis), they lay eggs that are never fertilized. However, because whiptails evolved from lizards with two sexes, they form couples in which they take turns acting as males by mounting each other imitating sexual intercourse. Apparently, this activity helps stimulate greater egg production.

What happens to the males? For some unexplained reason, eggs containing males are aborted before they hatch; only females eggs fully develop and are hatched.

➾**FYI:** Parthenogenesis also takes place among three species of the **rock lizard**. Although this is an efficient means of reproduction—since it requires only one being—it is also dangerous to a species. Sexual intercourse insures a varied genetic combination in offspring, making it difficult for disease to succesfully wipe out such a species. But with parthenogenesis, since all the animals have the same genetic code, an aggressive disease could make them extinct.

6
INCEST

Field notes. . .

Incest among humans is one of our strictest taboos, yet it is also widely practiced, the extent to which we have only recently become aware of. Generally, nature avoids incest because it reduces the genetic pool and results in weaker offspring with less chance of survival. This has proven to be a problem with endangered species. In some cases, there are so few of them that even when they do mate, they are dipping into their own genetic pool. The results are often fragile offspring. This is what's happening in Florida with the panther. Less than fifty are still left in the wild and their incestuous mating has produced offspring with congenital heart problems. Ninety percent of the males have one undescended testicle, which means a lower sperm count. To combat this, biologists have released female cougars from Texas into three southern Florida parks. Cougars and panthers are of the same species.

But sometimes incest is necessary for survival. Certain female animals will mate with males as long as there are males around. But if there are no males around, they will give birth to a batch of males and males only. Then the mothers will mate with their sons.

BAGWORM MOTHS

In 1995, the Government Accounting Office issued a report that by the year 2000 there would be more single-parent families than traditional two-parent opposite sex families. Certainly divorce is the main reason. But another contributing factor that is growing in popularity is that some women, frustrated in their search for the perfect mate, have decided to skip the husband (or nonmarried equivalent) stage and go straight to motherhood. Artificial insemination gives a woman an offspring without the pressure of having to select her mate under the deadline of a biological clock.

The same thing can happen among other animals. In some species of bagworm moth the female will mate with the male and produce offspring the old-fashioned way, but if there are no males around and she simply wills herself to give birth. This is a prime example of virgin birth (parthenogenesis). However, because too much of this kind of thing can weaken the species and make it vulnerable to parasites, the bagworm moth prefers the intercourse method of reproduction. So, when the female gives birth as a virgin, she gives birth to all males. Then she mates with them. Certain species of **leaf insects** also practice this method.

➾**FYI**: How does she produce only males? The egg of the female bagworm moth contains two nuclei instead of the one, something like a double-yoked chicken egg. Usually the two combine into one and grow into a larva. But with parthenogenesis, when she doesn't have a male fertilize the egg, she utilizes the male part of the egg to fertilize, thereby using her own egg as the surrogate "father."

GIBBONS

The gibbon is one of two species of ape that is monogamous and mates for life (see MARRIAGE: GIBBONS). They even take care of old grandpa (and grandma) when he becomes too old to remain as head of the family. So far, it sounds like the perfect Walton family. However, this family has a few peculiar habits that humans do not accept among themselves (though in the past they were perfectly acceptable among various peoples).

First, despite her marital fidelity, the female doesn't come into her first marriage a virgin; she has already had sex with her father first, and usually has given birth to his offspring. This first offspring is like the dolls little girls play with. It is a practice baby, with which the female learns how to care for an offspring. It usually dies from the mother's mistakes, but now she is ready to care for the offspring she will have with her true husband.

Second, the son also has some special duties in this family. When the male head of the gibbon family retires, his son takes over as head of the household. This includes "marrying" his own older mother. When she becomes too old, she is replaced by her own daughter. Thus, the brother and sister are now husband and wife. And so the family continues to renew itself by replacing each family "job" with another family member.

➾**FYI**: Though inbreeding is common, it is not the only type of marriage they form. Neighboring gibbons families will confront each other at the borders of their territories. They usually won't fight, but rather cry out at each other for half an hour or so. After they quiet down, they just stare menacingly at each other. But their young are allowed to mingle and play together. During this playtime, some of the young form relationships and wander off to start their own families.

GREEN BONELLIAS

The green bonellia is a marine worm that looks like a dill pickle with a yard-long rubber tube sticking out, ending in what appears to be two leaves. Innocent enough. Except it has a magical power like Circe's. Where Circe could turn men into swine, the touch of the female green bonellia turns neuter larva into males. And into her own husbands. When she gives birth to these tiny larva they are neuter; they float around with the current. But if they touch her proboscis (the pickle-shaped area), they cling there for several days, at the end of which they are males. They also stop growing, which leaves them stunted at a few millimeters, the size of a small bug (remember, the mother is about a yard long). Now the real work begins. This male eventually makes his way into his mother's intestines and into the oviduct, where he will spend the rest of his life fertilizing his mother/mate's eggs like a worker on an assembly line.

This work may not be as lonely as it seems, for researchers have found as many as eighty-five of these tiny males inside a single female. What happens to those ambivalent larva who drift through the sea like the Flying Dutchman without ever touching a female? If they are still mateless after a year or so, they simply begin to transform themselves into the treasure they have been looking for: they become females.

MOTH MITES

At first, one might find the moth mite remarkable simply because the mother's male offspring actually assist her in the vaginal delivery of their sisters. Which they do standing on their heads! But that's not the remarkable part. The moth mite blows apart any comfy notions humans have of family values.

First, some background: Moth mites are parasites, related to spiders and ticks, that feed off certain types of caterpillars.

The pregnant female finds a caterpillar to sustain her, then gives birth to live offspring that are fully sexually developed. When she gives birth to males, they bore into their mother's flesh and begin to feed off her juices. They could leave; they are quite capable of survival on their own. But they don't because they are patiently waiting. For sisters.

The mother soon goes into labor again. When she does, the males worm back out of their mother's body and wait in a group around the vagina like paparazzi at a movie premiere. If a male head appears, no one cares, and he is born without assistance. However, the moment the head of a female appears, the brothers rush to help, shoving each other until one of them is triumphant. He will then stand on his head and use his hind legs to pull her from the mother's vagina. The moment he has yanked her free, he mounts her from behind, thrusting his barbed penis into her vagina. She is born, raped, and inseminated, all within a minute or two. Now she is on her own, booted out of the family. She must find her own caterpillar host to feed on within a couple days or she will starve to death. Her brothers, though, remain with their mother, feeding off her juices and impregnating any more sisters that appear. (For similar behavior, see also SEX/DANGEROUS: BEE.)

➾**FYI:** A similar situation occurs with some types of **wasp**. The males develop more rapidly in their eggs than females and therefore are born before the females. Among these males, the one born first is stronger than his newly born brothers and is therefore able to drive them all off. Only the firstborn male remains to keep watch over the rest of the eggs. As each female comes out of her egg, he immediately mounts her and copulates. He will do this with each and every one of his sisters, no matter how many are born.

7
MARRIAGE

Field notes. . .

There are three basic types of marriages among animals:

(1) The "Juicy Fruit" Marriage. Two animals court, have sex, and go their separate ways. Their relationship is like a piece of Juicy Fruit gum: big burst of pleasure at first, but as soon as it loses its flavor, it gets spit out. Neither sex probably makes a connection between the sexual act and having offspring, they simply do what they are physically compelled to do. Once that compulsion is over, they see no need to stay together. In some cases, the female may specifically drive the male away, as among bird species in which the male is brightly colored. His raiment is a lure to predators who might threaten the offspring. This group would include those animals that maintain harems. Although some males with harems may be very tender and affectionate lovers, their bond with the females rarely lasts more than a few days or months.

Ironically, humans who equate marriage with the ideals of selflessness, might consider this type of temporary marriage as the lowest form of relationship because it appears to lack any commitment. However, just the opposite is true. Marital fidelity exists among aggressive species and seems to directly emanate from their sense of possessiveness. Among some animals, this form of temporary relationship exists because they lack possessiveness about nests and territory. Therefore, they are less quarrelsome, which means they don't have these extravagant lists of qualities their mate must possess. Because they get along with nearly all of their group, sex with one is pretty much the same as with another. It serves the procreative ends. But marital fidelity, as conceived by humans, demonstrates lack of commitment to the group and a more selfish need. Actually, this behavior reflects the same attitude expressed by many major religions as well: Christianity teaches that people must be able to form relationships with community that go beyond and are more important than their immediate families. Buddhism, Hinduism, Jainism, etc. teach that there are two levels of follower: the householder—the one who marries and raises a family—and the aesthetic, who is further on the road to salvation for having surpassed those relationships that keep people tethered to the material world.

(2) The "Wait-Till-the-Kids-Are-Grown" Marriage. Some animals form marriages for the sole purpose of raising offspring. They are often intensely devoted parents who cater to their young's every needs. The marital relationships among some such species appear to be close, but among most it is in fact distant and formal. Once their offspring have grown enough to be on their own—which can take a few months or years—these animals "divorce" and go their own ways. During the next mating season they will find another mate and begin the process anew.

Among this group are those animals that are married to the territory rather than the individual. This subcategory deals with marriages that are temporary, but not necessarily because of the offspring. Many species of animals form marriages that have nothing to do with the mate; rather, the marriage is between each animal and the property. The individuals involved in the marriage may have no "feelings" toward each other at all, but stay together because they inhabit the same space. (This is similar to the premise of Tama Janowitz's story "Slaves of New York," in which the characters, all residents of New York City, remain in bad relationships simply because it's easier to stay than try to find affordable housing in the city.) In such relationships, the individuals have no emotional investment in their mate, and therefore do not suffer if the mate dies or is replaced after battle by a new mate. In fact, they may not even be able to recognize their mates when they are away from the property. This kind of property-based marriage is practiced by various types of insects, fish, reptiles, birds, and mammals.

(3) The "Till-Death-Us-Do-Part" Marriage. This is the ideal among many—though not all—human societies: a monogamous, lifelong marriage of mutual devotion and love. Yet, it may be our ideal more in the breach than in the practice. We seem to like the concept of this type of marriage, but it is very rare. In the end, we judge a species by its actions, not just its aspirations. The high rates of adultery and divorce among humans suggest that our image of marriage is more romantic than real. Though humans assume monogamy is what we are evolving toward, at least "spiritually" if not biologically, some species of animal have actually evolved away from monogamy. More likely, monogamy, polygamy, or polyandry are tools from the same box, each to be used whenever circumstances require one over the other. For Nature, the goal is procreation and survival of the species—whatever that takes.

This kind of monogamous lifelong marriage generally requires an aggressive, possessive species. Such creatures are concerned about their nests, homes, property. They often have to defend these possessions with violence, hence their aggressive nature. This is especially true among species in which only those with property will ever mate. To pass along one's genes, one has to fight. This, of course, produces offspring that are more aggressive, since those are the only parents mating. Such an aggressive and violent daily life makes it difficult for an animal to suddenly turn into a softy just to go courting. Males and females alike don't trust each other because of their aggressive nature. So they have elaborate formal courtships that are meant to demonstrate that they will not attack or steal each other's food. This courting can be time-consuming, so some species find it easier to stay with one mate rather than court each season again and again. Monogamy is more efficient in terms of producing offspring.

These couples tend to stay together for life, even after the offspring have gone. They demonstrate affection for each other, and some display grief if the mate dies. An example of perhaps the most faithful married couples in the animal world were the **huia** birds of New Zealand. Their fidelity was a matter of necessity more than romance: the males had short beaks needed to strip away bark from tree trunks; the females had long, thin beaks that curved downward so that they could reach the larvae. Neither sex could get the food without the help of the other. Hunters made these birds extinct in 1907.

Among those who practice lifelong monogamy are badgers, some species of crab, cichlid fish, butterfly fish, gibbons, marmosets, many types of songbirds, ravens, doves, greylag geese, parrots, beavers, jackals, dwarf antelopes, and some whales. But there are many variations of behavior even among the monogamous. Also, lifelong monogamous marriages do not exclude incidences of adultery.

The "Juicy Fruit" Marriage

BABOONS

During mating season, the female **anubis baboon** of Kenya mates every day—each time with a different partner. This is not the random sex of a too-drunk debutante at a frat party. She selects her mates in a deliberate order starting with the lowest-ranking male in her group and working her way up to the highest-ranking male. It is at this later time that she is most susceptible to becoming pregnant, so she has timed her sexual encounter with the head male to coincide with when he will most likely be the father. Taking additional lovers confuses the paternity of the child, thus assuring that she and her offspring will be treated well by all the males, since they can't be certain who the father really is. Thus, "marriage" is to the group rather than an individual male.

Geladas and **hamadryas baboons** of Ethiopia and Somalia have a different social system. Males maintain small harems of two or three females who are permitted to mate only with their master. Any female who wanders too far from the other members of the harem is rounded up and punished by the male with beatings and bites.

This difference in marriage structure between the two species of baboon reflects the influence of geography rather than personal preference. The geladas and hamadryas baboons live in harems because they inhabit barren terrains that have a very small food supply. They are forced to search large areas for enough food to feed even a small harem. If they traveled in groups, they would not survive. However, the anubis baboons, which live in groups of eighty or so, inhabit large fertile plains that provide plenty of food. Unfortunately, this also means plenty of predators, so they band together in a large group, the better to defend themselves against lions, hyenas, and others. Small harems would be wiped out.

➾**FYI:** Establishing a harem is not a simple matter. One method involves kidnapping or eloping with the young daughters of females from another harem. It works pretty much the way it does among humans: As daughters mature, the mothers treat them like sexual rivals, punishing them more harshly than required and generally making their lives miserable. This is the perfect time for a young male baboon to sweep in and snatch the unhappy female away. For a brief time, the male does not attempt to have sex with her, but rather behaves like a surrogate parent, nurturing her emotional dependency. When he does have sex, she is now part of his harem and any disagreements are met with a smack.

The second method is to become a lieutenant male in an established harem of an older male. The male of a harem becomes less aggressive as he ages, and therefore more tolerant of the presence of other males. If a younger male approaches in the proper manner—displaying submissiveness—he may be allowed to hang around and help out with some chores, such as keeping the females from straying. With time, he gains more and more responsibility, deciding where to hunt for food, when and where to sleep, etc. But he still is not permitted any sexual privileges; that must wait until the prime male dies, at which time the lieutenant inherits the harem.

CHIMPANZEES

Chimps live in a community of adult males and females and offspring. They travel together through the forest, living in trees and on the ground, though they always sleep in the trees. Each chimp sleeps alone, except for the very young offspring who sleep with their mothers. They roam through an area about ten miles square, rarely sleeping for more than one night in the same place.

The community is further divided into single-parent families, with the mother being the head of her household of offspring. Males do not belong to these smaller family units because chimps are promiscuous and no one knows who fathered which offspring. In general, the males ignore the females until one of them comes into estrus. Then the skin around her genital area becomes swollen and bright pink and she emits a sexual scent, which the males quickly discover. Suddenly she becomes Ms. Popularity.

GROUSES

The **black grouse** is a perfect example of the "Chippendale's Syndrome." The males get together and perform an elaborate display for the females. They leap, they dance, they sing, they mock duel, they show off their lovely feathers—all for the entertainment of the females. When one of the plain females selects a hunk of burning grouse, she stoops and invites him to mate. Once he does, she gives him the instant Dear John response. Because his bright, showy appearance calls attention to predators, he is now a threat to her and the young she will have.

Even more spectacular looking is the **sage grouse**, common to the North American prairie. About four hundred of these males will gather in a courtship area about eight hundred yards by two hundred yards, each staking out a parcel of land for himself. At the center are the four BMOCs—these are the guys the hens most want to mate with. In fact, 74 percent of the hens will head straight for these fellas; when the Big Four are too tired to satisfy them, a handful of second-ranking cocks will step in to take up the slack, which is about 13 percent of the hens. Another 13 percent mate with lower-ranking males. That only totals up to about fifty males out of four hundred who get to mate; the rest will probably never mate. Once the mating frenzy is over, the males ignore the hens, who wander off to build nests and raise their offspring.

➾**FYI 1**: The **hazel grouse**, a close relative of the black grouse, leads a very different life. Both males and females are very plain looking. During the courtship period, the male goes through similar display rituals as the black grouse, but on a much more low-key level. His one bright feature is the red combs that he can cause to peek through the blander feathers. Apparently this is enough to get a mate, though they don't actually have sex until six months later. And once they do mate, because the hubby resumes his plain looks and calm demeanor, he presents no threat to the female or their young. They remain married for life.

➾**FYI 2**: The male **ptarmigan** is another bird with a coat of many colors. His bright feathers attract a female with whom to mate, but make him a dangerous companion with whom to stay. Apparently, it's not enough for the male to be the happy bachelor, flexing his feathers and pecking notches on his perch for each conquest. At heart he's a family man, despite his appearance. So, after he has mated with a female, he watches over her and the chicks from a distance. While she watches over the offspring, he stands ready to protect them or distract approaching predators. In autumn, he grows a whole new set of feathers that make him as drab-looking as his wife. Now he moves back home with his family. (In winter, male and female will both grow white feathers so they are camouflaged by the snow.)

GNUS

The **brindled gnu** of the Ngorongoro Crater of East Africa would make Karl Marx's beard knot. Among them, as is true among many animals, only the landowners have sex. The unlanded peasant males roam in herds, indifferent to each other, waiting only for their opportunity to seize some land and finally mate. This is called the "cattle system." In fact, male gnus never fight over females; they only fight over the land itself, for it is the land that attracts the females. This system of a bunch of gnus defending their individual patches of land doesn't allow for very close social bonding. The only contact males and females have is during the few minutes they mate. So no marriage as such is ever formed. (For similar behavior, see COURTSHIP: KOB.)

However, not all brindled gnus adhere to this landowner/peasant arrangement. Those living outside the Ngorongoro Crater are wanderers. During the mating season of May and June, they interrupt their migratory travels while the males stake out temporary territories. A herd of bulls will corral a herd of females and divide them up with a couple dozen females per bull. After the males have mated with their harem—which takes a couple hours—all return to the herd to continue their migration.

➾**FYI 1:** Many species of **gazelle** and **antelope** also adhere to the practice of only landowners being allowed to mate. As a rule of thumb, the more land a male possesses, the worse he treats the females. The male **Defassa waterbuck** and the male **impala** are ever vigilant if any females try to escape from their property, herding them back to the center where they remain imprisoned. But those animals with very little land, such as the **Uganda kob**, who owns a measly ten to thirty yards of land, must treat the female with much more kindness or she'll just bolt the short distance to the neighbor. But the more land the male controls, the more abusive he can be because he has plenty of room to chase her down if she tries to escape.

Whether large landowners or guys with little yards, whether permanent or temporary, territorial mating systems do not encourage marriages to be formed since there's little opportunity for social bonding. Therefore, this system is designed for mating purposes alone—that the strongest and therefore those with the best potential as gene donors have an opportunity to reveal themselves. Once the sex is over, they go their separate ways and the females raise the offspring as single parents.

➾**FYI 2:** It's a bachelor's dream society: fifty to five hundred males roaming freely, without any social hierarchy. No one's in charge, no one outranks any other. No one picks on another, rarely is there any fighting. That's how many of the brindled gnus of East Africa live. However, there's a down side. This ain't *West Side Story* ("When you're a gnu, when the spit hits the fan/you've got brothers around, you're a family man."); these herds don't protect each other from predators nor do anything to help each other. They just happen to be traveling companions—nothing more. Their real goal is to find a plot of land they can claim for their own so they can finally mate.

HORSES

A male horse (stallion) generally has a harem of female horses (mares) that he protects as well as impregnates. When the mares come into heat (every twenty-one days), they rarely offer resistance to his advances. While in heat, the mare may urinate, then deliberately step into it; this is why you will often see a stallion sniffing at the ankles of mares. The stallion may respond to the scent by wrinkling his nose and pulling back his lips to expose his teeth, a gesture common among plant eaters. Oddly, the female may actually come into heat even if she hasn't released an egg from her ovary, or she may release an egg and not come into heat. This often confounds the purpose of coming into heat, for the horses may engage in sex without any possibility of pregnancy resulting.

➠**FYI:** Male **donkeys** (jacks) do not share the same easy access to their females (jennies) as stallions do to mares. Male donkeys have to chase the females and violently battle them in order to force them to stand still long enough for them to mount. Like some other animals, the female cannot copulate without violence. When the male donkey happens across a willing mare, he is often startled by her submissiveness. The stallion is equally as startled by the combativeness of a female donkey he may desire to mate with. This is why most mules (which are offspring of a horse and a donkey) are usually the product of mating between a male donkey and a female horse.

POLAR BEARS

Coca-Cola made the smiling polar bear an icon of the happy Coke drinker. Smiling furry white stuffed polar bears holding a Coke are sold everywhere. Real polar bears probably don't drink Coke, and they definitely don't smile. In fact, they don't show many facial expressions because their foreheads and cheeks contain no muscles. This makes it difficult to express emotions to each other, which may contribute to their solitary lifestyle. [The inability to communicate emotions, or the repression of emotions, is also one of the main causes of human divorces. However, humans have the ability, both facially and verbally, to communicate; when they fail to (as men are often accused of doing), it is often interpreted as neglect.]

When a couple of eight-foot tall, fourteen hundred pound polar bears finally do run across each other and decide to mate, a certain amount of communication takes place—in the form of wrestling. They stand on their hind legs and start shoving each other; eventually, shoving turns into affection. The marriage lasts two or three weeks before they call it quits and move on. It is unlikely the same couple will ever mate again. Next mating season they will find new spouses.

SEALS

For those who have ever fantasized about the pleasures of having a harem, let the harried life of the male **northern fur seal** serve as a grim warning. First, we should get an idea of the behavior of the seal before he's assembled his harem. During the fall and winter migratory season male seals are sociable creatures, intermingling with each other as well as with seals of other species. But come spring and summer mating season they're like sailors who've just had their shore leave canceled. They all return to their place of birth in May, arriving a full month ahead of the females. Here on their island, they choose a hunk of beachfront property as their own and defend it against all other males. Just a short swim ago these were playful, friendly creatures; now they are mean and hostile as they stand guard daily over their land.

In June the females arrive and the males get even worse. They fight and bully each other as they aggressively go about gathering their own harems, which range in size from two to a hundred, though thirty is about average. The bigger and stronger the bull seal, the larger the harem; this ensures that more of the herd's offspring will be stronger. Once he's rounded up his harem, the male is not yet ready to take advantage of them. This is because they are still pregnant from last year's mating season. The females return to the island, not just to mate, but to give birth. Usually within forty-eight hours of arriving, a female will deliver her offspring; she will be able to mate again a week later.

But this week can be a long one for the male, who now must

not only fight off all other male usurpers, he must keep guard over the females to prevent them from deliberately sneaking off for sex with another seal. The female is more a prisoner of war than a mate; she is treated as property and without any affection. With all the fighting for land, defending of property, gathering and guarding of the harem, the male hasn't eaten for a month (losing as much as sixty pounds of muscle and fat). And he won't eat until after he's had sex with all of his harem. This exhausting schedule weakens males to the point that their enemies—bachelor bull seals that have been exiled to the cheap beach property—may decide to attack. Sometimes the young usurper succeeds in deposing the older bull and stealing his harem.

Even the most hearty of bulls only has a three-year reign as harem head. That's when he is strongest and best able to round up the females and defend his territory. After that, he's like the star of a popular sitcom that got canceled: he keeps trying to regain his previous popularity. Forty percent of the males between seven and seventeen die each year from bite wounds they received in combat with other males. This leaves a disproportionate number of females to males, hence the necessity of the harem system.

➾**FYI:** Female **sea lions** have a similar system of swimming ashore to give birth and then joining a harem. The difference is that the females select their mates and stay with them out of choice, not intimidation. However, a female sea lion only stays with the male—who is twice her size—if he is kind to her and courts her. In fact, he must court her for at least a full day before she will slip into the water with him for mating (fur seals must mate on the hard rocky land so the male can keep one eye on his harem who would try to escape). Like the fur seal, the male sea lion must defend his harem against other males who would try to steal them, but unlike the seal, he doesn't have to guard the females to keep them from waddling away. Females stay with their mate—unless he gets deposed—because they've chosen him. And the females of the harem may have a hand in that coup.

Mating takes place in the water. The couple may float about side-by-side for over an hour; but once the sex is done, they separate. He may doze off in the water while she returns to tend her newly born calf. Within minutes another of his harem will impatiently wake him, insisting on her conjugal rights. And so it goes, with the male losing weight and sleep until his sexual energy is depleted. This might result in the females crying out in unison (they get along quite well together), which in turn encourages nearby bachelors to hustle over to try to defeat the weary male and take his harem. Each year a harem may change males several times. Though it may seem as if the male is the head of the harem, in fact he may be more like a hired hand, easily replaced if his workmanship slips below expectations.

The "Wait-Till-the-Kids-Are-Grown" Marriage

ALBATROSS

Separate vacations to rekindle the old spark? Works for the albatross, which breeds only every two years. First, the male arrives at the old nest about a week before the female. When she arrives they go through the old courtship ritual, mate, raise their single offspring, and live together in harmony for six months. After that, they fly off in their separate directions for a year and a half. When they return, it is to the same nest and therefore to the same mate—which is why the albatross mates for life.

➾**FYI:** The albatross marriage is dedicated to one cause: the upbringing of their young. The parents take turns guarding the nest; while one parent flies out to sea to fatten itself for several days, the other remains at home starving. When the fattened mate returns, the other flies off for several days of chowing down. However, both parents will feed the offspring, which will grow fatter and bigger than the parents. Finally, the parents will desert their fledgling and fly off to sea to molt. Still in its nest and too fat to fly, the young albatross begins to starve. The length of this starvation depends on the type of albatross, but it is usually at least a week and in larger species much longer. Eventually, the chubby bird slims down enough to be able to fly and gather its own food.

BLIND GOBIES

The more choices we have, the more selective we seem to be. The more potential mates that are available to us, the more exacting our list of qualities of the perfect mate. But imagine a person of the sex you prefer—except this person is everything you hate, every loutish mannerism, every disgusting habit. No way the two of you could ever be together. Then imagine you are stranded on a remote island with that person, for years on end with no hope of rescue. It is quite probable that love will blossom—not because of the innate qualities of that person, but because of our need to love.

The blind goby fish is in that same situation and demonstrates the "marriage to territory" discussed above. The goby is a loner, living as a guest of the tropical mole crab who digs a cave, then spends most of his time keeping the sand out. Meanwhile, the goby just hangs out eating the microscopic animals that wash through the cave. (Unfortunately, if the crab dies, the goby must find another crab with room, or it too will die.) Despite the fact that they do nothing to dig or maintain the dwelling, male gobies are very aggressive and do not allow other males into their homes. However, they will not attack females. So, if a female happens to take up residence in the same cave, they mate. His relationship with her, and hers with him, is based on proximity, not attraction or emotional bonding. If she dies and another female swims in, she becomes his mate.

STORKS

Storks return from migration each year in search of their previous year's nests. The males arrive first to reclaim the nests and await the arrival of the females. It was once assumed that storks returned seeking their same mates. But now it is understood that they seek the nest instead, another example of "marriage to territory." The females, too, want their old nests back; the males occupying them are incidental to the nest itself. If a young female comes to flirt with the male occupant of the nest, the female will attack her and try to drive her away; the male may do the same with a flirtatious young male dancing near his nest mate. However, these displays of jealousy are not related to the mate but to protecting the nest. If a female fails to drive away the younger female and is herself banished by the newcomer, then the new mate moves into the nest and is accepted by the male in residence; the female will do the same if the male is driven off.

The "Till-Death-Us-Do-Part" Marriage

BEARDED TITS

Sometimes an animal can invest so much emotion in one relationship that if something ever happens to their mate, they are never able to feel as strongly for another. Humans are forever making declarations like this to their mates, though we seem to be much more resilient than we give ourselves credit for. In fact, the first "great love" may be so filled with passion as to drive us crazy with the chaos of it. When it's over and we find another mate, the new relationship may be much more satisfying because of what we've learned from that first romance.

The bearded tit is not so resilient. When these European birds form a relationship, it is for life. At night, the male encircles his mate with his wings to keep her warm. If a bearded tit loses its mate, it suffers through a long period of mourning and depression. Finally, it may again choose another mate. However, adult birds, which are calm and steady, have difficulty forming a bond with the younger birds, which are noisy and aggressive. These May-December relationships require adjustments and this second "marriage" is therefore rarely as loving as the first. Sure, they produce offspring, but the couple tends to quarrel more and spend more time apart than did the first couple. If the arguing becomes too frequent, they may divorce and go their separate ways.

BULLFINCHES

"Till death do us part." Nice sentiment among humans, but many worry more about disability than death, the ole "in sickness and in health" part. Death is easy, but will one's mate stay and see one through debilitating illness? Among bullfinches the answer seems to be yes. When both are healthy, the male and female take turns sitting on the eggs, though when the female is on the job, the male hustles about feeding her. But if he gets sick before the eggs hatch, he takes to the nest and she goes out foraging for food. When she returns, he opens his beak and acts just like a baby finch, triggering her maternal instinct. Though most sick birds will die unless humans intervene, sometimes they get better. Yet, even if he doesn't, the female's nursing of him will probably extend his life long enough for him to tend the eggs until they hatch. It is unknown whether her behavior is a form of affection or just practical parenting.

➾**FYI:** Though bullfinches are monogamous and do mate for life, they are not without a checkered past. Long before they meet their mate, they have been engaged before—*to their own brother or sister*. When the half-dozen siblings reach six or seven weeks of age, they begin to court each other. They are still a year away from being able to engage in sex, so this flirtation is more a practice run that teaches them proper behavior toward the opposite sex. They treat each other with affection and, whether male or female, all behave like females. They offer the other the opportunity to mount, though none are yet capable of accepting the invitation. Still, they eat and sleep together, defend each other, and generally act like a loving married couple. But as they mature, they begin to have fights until they finally are old enough to fly off on their own and find a mate outside the immediate family. In fact, their aggressive attitude toward their former "mate" remains should they see each other again. This prevents incest.

The second engagement—which is to a bullfinch outside the family—is based solely on compatibility. If they click, they remain together; if not they separate and keep looking.

BUTTERFLY FISH

The *Chelmon rostratus*, a type of butterfly fish that lives in the coral reefs of the Indian and Pacific Oceans, is truly monogamous in that it mates for life. This may make them an ideal romantic couple in human eyes, but as parents they lack the same level of commitment. Once they have spawned, they pretty much ignore the offspring and swim off together. Part of the reason is that this species tends to be an unsociable lot, even aggressive with others of their kind. By staying "married" to one female, the male avoids having to go through dangerous combat with other males each mating season.

CARP WORMS

The carp worm lives permanently on the gills of certain species of carp. When they take time off from their underwater travels to engage in sex, things get complicated: each carp worm possesses both male and female genitalia. That in itself is not that unusual (see EARTHWORMS, among others), and to have sex they simply have to match each penis to the other's vagina. What is unusual, however, is that once they have joined together, they never come asunder. That is, they remain in the same position for the rest of their lives, locked in sexual embrace. (See also ANGLER FISH for a similar lifetime "commitment.")

GUANAY CORMORANTS

These birds of Chile live in giant colonies of hundreds of thousands, yet remain monogamous to their mates. Their courtship consists mostly of the male sitting in the nesting site while the female flies special flight patterns. Once he allows her to land, they continue the getting-to-know-you ritual by twisting their necks back and forth. Either one can cut the date short and move on to someone else. But if they don't, they begin to run their beaks through each other's feathers in a caress. Afterwards, the male flies off to fish; if the female is still there when he returns, they're married.

GIBBONS

In general, monkeys and apes are not monogamous. The gibbons and **marmosets** that live in the jungles of South

America and southeast Asia are the only apes that practice monogamy within a family structure of two parents and their offspring. The **siamang**, a close relative to the gibbon, also is monogamous and mates for life. All are very aggressive species, especially toward members of their own sex: the males chase males, the females chase females. But neither sex is aggressive with the opposite sex. Marriage relationships are quite cordial. Neither mate tries to take food from the other, and they hug and caress a lot. The one complication concerning the gibbons' monogamous marriages is that they practice generational incest (see also INCEST and PARENTING).

➾**FYI:** Most primate societies, however, practice promiscuity and polygamy. But unlike some other species, including humans, that are promiscuous, jealousy is rare among primates. The usual pattern of polygamous behavior involves the female in heat mating with pretty much any male in the group. In some species, she may have sex with every male in the group, but does so in a methodical way, starting with the lowest-ranking male and working her way up to the leader. The reason for this pattern is that her eggs don't descend into the uterus until the latter part of her estrus; it is at this point that she is most likely to become pregnant, so it is at this point she ends up with the top male.

HORNBILL

Protecting one's family from intruders often forces fearful families into the siege mentality, constructing bars on their windows, roll-down steel shutters, and other devices that sometimes makes the people inside feel like prisoners. The hornbill, a southeast Asian bird roughly the size of a turkey, creates the ultimate home-protection system. When it's time for his mate to lay her eggs, the couple finds a roomy hollow in a tree. The female slips inside and together they begin constructing a wall to close off the hollow. Because the male's on the outside, he uses mud, saliva, and resin as building material; because she's confined to the inside, the female uses her own excrement. When they are done, the hollow is "cemented" over except for a small feeding hole, large enough for her to stick out her beak. This is where she will stay for the next two months, walled-in inside her closet while hatching the eggs. This wall keeps predators like apes and snakes away from the eggs. Outside, the male spends most of every day hunting and feeding his family, sometimes delivering as many as sixty pieces of fruit each visit. After about two months, when the chicks are old enough, the mother pecks the wall apart and rejoins her mate.

Technically, the female could break out the wall if she wanted. But there is a bigger problem than the wall. Once the female is enclosed in her little room, she molts, losing all her feathers—which she uses to line the nest—and thus her ability to fly. If her mate is killed, she and her young are in serious danger of starvation. That's why the bachelor hornbills patrol the area during mating season like some sort of Neighborhood Crime Watch. They spy on the couples, looking for any female who might be neglected. When a bachelor male finds such a female, he assumes the mate is dead, for he would never intrude in an established domestic relationship. But a widow is fair game, and he instantly assumes the role of provider, feeding the female as his own mate, which she now is even though they have never had sex and the offspring are not his own. A marriage relationship has been established separate from the sexual drive. After the young have left the nest, the couple remains together for life. However, some males are so weary and gaunt after their season of caring, a sudden turn in the weather may kill them.

➾**FYI:** Some African tribes copied the hornbill's methods, though completely misunderstanding its purpose. These tribes would imprison their wives throughout their entire lives.

LOVEBIRDS

Lovebirds establish social ranking by sitting side by side on a tree branch. Without warning, one will peck at the other's foot, grabbing the foot and yanking, causing the rival to flip off the branch. However, females do not necessarily choose their mates merely by social ranking; apparently there are other qualities that they weigh. Whatever they are, once a couple chooses each other, they remain mates for life. If the mate dies, the survivor will remarry, preferably a former suitor or suitee.

Lovebirds get their name from their technique of passing food to each other on the tips of their tongues. To do this, they tilt their heads slightly so they resemble humans kissing. But they also do this kissing after periods of even short separation, when they are frightened, even after a disagreement. Only mates may kiss each other.

Yet, what appears as an ideal marriage to humans can change abruptly with conditions. Behavioral experiments have shown that if there are more females than males, the females begin courting males, even the married ones. A wife will battle her rivals, but if she loses, the male may suddenly have a harem of several wives. In such cases, the females will raise their young by the same male in separate nests. However, if there are more males than females, the females do not suddenly take on additional husbands—at least not sexually. Those males without mates may hang around with married females and may even feed the females, but that is where the line is drawn. They may not kiss a married female. However, if the male mate dies, one of these "friends of the family" may then take over as the new mate.

➾**FYI:** The **budgerigar**, a close relative to the lovebird, also mates for life. But both husband and wife (especially the husband) commit adultery, though these extramarital affairs do not affect their apparent devotion to their personal relationship. The result of these infidelities is that the male seldom raises offspring he fathered, though he is unaware of that fact.

An even more interesting variation of their family arrangements is that, because their personal and sexual relationships are treated as separate, budgerigars of the same sex will sometimes form a permanent "marriage," though without the sex. They remain lifelong companions. Two female nest mates will have sex with different males, but then both will lay their eggs in the same nest and raise their young together.

MACAWS

Macaws are those gorgeously plumed birds that are so popular in pet shops. Their popularity has led to their numbers being severely depleted from their home jungles of South America. Ironically, when captured and chained to their perches they display much the same behavior as humans in a similar situation—they become extremely depressed. This is because, like humans, they are very sociable creatures and do not do well when removed from the company of other macaws. Another irony: Macaws' domestic relationships actually represent the ideal that many humans strive for; they mate for life, and since they live to be as old as seventy, their marriages can last for many decades.

Macaw marriages are not just long—they are also models of gentleness and commitment. Male and female not only mate and raise their young, but they defend each other against others. The male is also a generous gift giver, bringing his mate fruits and Brazil nuts (which he cracks open for her with his hard beak). He will also wrap her in his wings and hold her close. When she is hatching eggs or parenting their young, the male will bring back food for her, some of which she eats and the rest of which she uses to feed the young.

MICE

A female mouse is monogamous. She will breed only with the same mate. If another male copulates with a pregnant mouse, or even tries, she will abort within four days.

PENGUINS

Human fiction is filled with stories of lovers searching each other out, sometimes halfway around the world, convinced there is but one true love for them. The female **Adelie penguin** is no less determined. Winter in Antarctica begins in March, and the males are the first ones out of the freezing water. They hop ashore by the thousands, pacing around like frat boys waiting for the neighboring sorority to arrive. And arrive they do, but in April, when more of the sea has frozen over and the females now have to travel over miles of icy terrain in search of the males. But not just any male. Most of the females already have a "husband," though finding him among the very similar-looking males can be a challenge. She approaches each likely candidate, sings, and awaits the appropriate response. If it isn't the right response, she keeps searching. When the couple is finally reunited, they throw their heads back, stick out their flippers, and sing loudly to each other.

Stephen Stills used to sing, "If you can't be with the one you love, love the one you're with." The female penguin agrees. If she can't find her old spouse, she will accept a new one. However, if the old spouse finds her within the next couple days, she will abandon her new lover for the old one. For the next two weeks they will walk around, side by side, sometimes touching flippers or bills. After two weeks, they have sex, the desire for which is signaled by the male laying his head across her stomach. Together the two trundle off away from the colony in search of seclusion. The actual sex itself is a clumsy affair, with the male climbing astride her from the rear. The rounded shapes of their bodies makes this a difficult position, and the male can often lose his balance and fall off her. If this happens too often, the female will walk away. But when he is successful and manages penetration, it's all over within three minutes and the couple returns to the colony, not to mate again for the rest of the year.

➾**FYI**: Among penguins, as among many other animals, the more aggressive the species, the more faithful it is during marriage. Adelie penguins are one of the smallest yet most aggressive species of penguin—and one of the most faithful. The huge **emperor penguin** is peace-loving and laid back—and among the least faithful. One major reason is that Adelie penguin is more territorial; it builds a nest, complete with a stone wall around it. This territorialism (as well as the birds' constant thieving of stones from each other's walls to build their own wall) causes frequent fights. But once the females lay their eggs, the quarreling stops and they live relatively at peace with each other. **King penguins** do not build stone walls, but shelter their eggs by carrying them on their feet (see PARENTING). Also, they move around more and therefore have no specific territory to defend—and so no need to be aggressive. Being less possessive about land makes them less possessive about their relationships; they tend to return to their previous mates at a much lower rate than the Adelie penguins. The emperor penguin is a big clumsy animal with hardly any aggression. They are sociable, often huddling together by the hundreds to keep each other warm. Because they all get along so well, mating with one partner is no better or worse than mating with another. Therefore, they are even less faithful than the king penguin.

SHRIKES

The **woodchat shrike** have the kind of marriage that is common among humans—enough so to support a whole industry of marriage and family therapists. They seem to get along just fine while raising their young: They hunt together as a family; the parents remain monogamous. A real Brady Bunch bird. But once the young have gone their own way, tension mounts. Not right away, of course, because there's still plenty of food until September. As long as there is food, there is no reason to separate; their former physical bond keeps them together. Once autumn and winter hit, the food supply becomes scarce and the shrikes begin to bicker and fight. To avoid any further damage to their relationship, they separate. But not too far from each other. Though they don't actually come in contact, they stay close enough to make their presence known to the other, either through cries or by flying over the other's territory.

In March, as the new mating season approaches, they return to each other and live together once more as a couple—until the food gets scarce again.

SWANS

Swans who are attracted to each other bow their long necks out, then lean in to bring their foreheads close together. A side-view of this position looks like they are forming a Valentine heart. This signals that they are in essence engaged. Engagement precedes mating. Although they become engaged in December and January, they won't be sexually mature enough for mating for many more weeks. Once they do mate, they are monogamous, remaining faithful to each other for the rest of their lives. However, a partial explanation for this behavior may be that they breed in isolation and therefore do not have the opportunity to be unfaithful.

TERMITES

How many marriage counselors have heard the complaint, "Sex has become a chore!" Among termites, sex is indeed a chore—but only if you're lucky. A typical termite colony has a population of millions, but only two of them have sex. The king and queen, who are the only ones with wings, are also the only ones who have sex, which makes the rest of the millions their offspring. Producing offspring is their only job, and they are very good at it: The queen lays an egg almost every two seconds, which equals about eleven million offspring a year. She does this for about twelve years. The king waits at her side during the deliveries; the rest of the time he's busy fertilizing her.

They are well compensated for their efforts. They have an especially large room all to themselves and an endless stream of hundreds of servants who cater to their every need, including feeding, washing, and carrying away the eggs that the queen is constantly producing. And they have their own royal guard of soldier termites to protect them.

But, as with most creatures, time takes its toll. After a while, the svelte termite queen begins to lose her looks. First her wings go, then her figure. She begins to put on weight until she is so bloated that her mate has to dig his way under her enormous body to find her genitals in order to have sex. Nevertheless, he is as loyal to her as his subjects/children are to them; he remains with her.

➾**FYI:** The beehive is similar to the termite colony in that only two of the fifty thousand members of the population get to mate, the queen and the drone she deems worthy. But, unlike the king termite who continues to have sex with the queen for over a decade, the queen bee and the drone mate only once. That one time is all she needs; she remains fertilized for the rest of her life. Since she no longer needs the drone, she kills him after sex (see SEX/DANGEROUS).

8
MASTURBATION

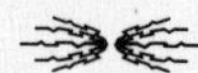

Field notes. . .

Most religions discourage masturbation as a symbol of "unnatural" behavior—unnatural because the function of sex is procreation, and any act of sex that does not further this natural purpose is mere indulgence and therefore perverts sex into a means of sensual gratification. Such teachings suggest that masturbation is merely a symptom of someone who is unable to control the impulse toward serving the desires of the self (devil) over the needs of the community (God).

Despite these teachings, statistics indicate it is a popular activity among humans. A *Kinsey Report* of Americans in the 1940s and 1950s concluded that 94 percent of males and 40 percent of females masturbated. The more recent and more comprehensive study, *Sex in America*, reports that among Americans aged eighteen to fifty-nine, 60 percent of the males and 40 percent of the females masturbated. However, among those living with a sexual partner, nearly 84 percent of men and 45 percent of females masturbated, indicating that masturbation isn't merely a substitute for unavailable sex. Still, despite how widespread the activity is, in December 1994, when

Surgeon General Joycelyn Elders suggested masturbation was a natural part of human sexuality and should be at least addressed in sex education classes, she was promptly fired by President Bill Clinton.

Is masturbation "natural"? Theologians are often conflicted by what is observed in nature regarding masturbation. For many centuries it was assumed animals did not masturbate and this was offered as proof that masturbation was unnatural. When it was discovered that many animals do indeed masturbate, this was offered as proof that animals are a lower form of life than humans since they cannot control their passions. Then it was theorized that those animals that masturbate do so only because they are in captivity, thus proving masturbation is the act of a damaged mind. Then it was discovered that some animals do it in the wild and they do it just because they like to. And that proved. . . oh, I forget.

Many animals in their natural habitat do masturbate. When a female reaches estrus, she is driven to copulate—her eggs are ready to be fertilized. She doesn't necessarily know what is going on in her body, only that she has a compelling need, an internal "itch." And if there is no male around to satisfy that itch, she scratches it herself. A female will rub her vulva against inanimate objects like trees or fallen branches, or she may drag her genitals across the ground. Or, if she is physically able, she will lick herself. Males, too, have their ways of soothing their needs, as anyone whose leg has been grabbed a little too tightly by Frisky the dog can attest.

DEER

Male deer use their antlers in deadly combat, gorging their opponents into submission or death. They use those same antlers to masturbate. The stag's sexual desire is so great that even his harem of females isn't enough to satiate him. He may then masturbate several times a day, which he does by rubbing his antler tips gently against shrubbery. It only takes about fifteen seconds of that for the stag to achieve an erection and ejaculation.

DOLPHINS

Humans seem to have developed more of an affinity for dolphins than any other creature, at least regarding their intelligence and humanlike qualities. Dolphins have also earned a reputation as being as sexually randy as humans, though a lot less inhibited. They engage in copulation even when there is no reproductive drive, apparently just for the fun of it. This enthusiasm for sex spans all ages and both sexes. Even though they don't reach sexual maturity until they are five years old, six-week-old males still participate in sexual intercourse. Dolphins also engage in homosexual activity. (See also HOMOSEXUALITY: DOLPHINS.)

➾**FYI**: Captive male dolphins masturbate with no inhibitions, even in front of the females. One method often used is for the male to situate himself in front of the jet stream shooting into the tank; the water rushing across his penis provides plenty of stimulation. The captive male dolphin's erect penis can also serve purposes other than the sexual; males sometimes use it like a blind person's cane to probe the bottom of a new tank.

LIONS

The male lion, despite an enormous amount of copulation (see also SEX/DURATION & FREQUENCY: LIONS), still sometimes finds the need to lie on his back and masturbate by rubbing his penis between his hind paws.

PORCUPINES

Sex between prickly porcupines is tricky (see SEX/DANGEROUS: PORCUPINES); perhaps that is why they spend so much time in self-stimulation. During mating season, females are anxious for sex, but as their desire grows, their vulva becomes less and less sensitive. As a result, they may straddle sticks (or anything else around) and rub themselves into ecstasy. When the male and female meet, should the female reject the male, he may straddle the same stick she just used and rub it against his penis and anal glands.

PRIMATES

Primates of all types have been observed to masturbate in the wild. However, this behavior is especially popular when they are confined to zoos, where masturbation seems to be a response to boredom. Although both sexes engage in masturbation, males have the ability to masturbate using their hands, feet, mouth, or tail.

9
PARENTING

Field notes. . .

A Japanese proverb states that "a father's goodness is higher than the mountains; a mother's goodness is deeper than the sea." Indeed, it is here, in the intimate and treacherous world of parenting, that humans see themselves at their best—and their worst. First-time parents, surprised to find that their newborn doesn't come with an owner's manual, scramble to bookstores and libraries for guides to usher them through the turmoil and tribulations of parenting. Psychology has made us tremble with awareness (however erroneously at times) that every harsh word or lack of harsh word can cause irreparable damage. But the national epidemic of child abuse, including shockingly high numbers of incest cases, indicates that being a parent doesn't guarantee a fraternal or maternal instinct.

How does one learn to become a good parent? Before the actual birth, soon-to-be parents concern themselves with how not to raise a neurotic child who will form a mutant religious cult, kill his followers, then blame his parents when he's interviewed by Barbara Walters. But when the baby is born, first-time parents are much more concerned with just keeping the little screamer alive. Fortunately, when people have their first child, they may be panicky about how to raise it, but they have the benefit of others' experiences, be it in the form of advice from parents, friends, or experts. Animals don't often have that extensive of an support system, so they have to learn it the old-fashioned way: trial and error. That's why some species actually have a practice offspring on which to learn. Gorillas do this, as do most bird species.

Parents' self-sacrifice for their offspring reaches amazing heights; conversely, their apparent cruelty can plunge to terrifying depths. Parents will sacrifice their lives for their offspring, yet those same parents will also cannibalize their own young. But both acts serve the same purpose: they are necessary to insure survival of the species.

AMOEBAS

Sex and death are inevitably linked in literature, philosophy, and religion. In *The Catcher in the Rye*, Holden Caulfield is terrified by his own puberty. He believes that sexuality is the beginning of phoniness and corruption, which signal our decay into death. In more practical terms, an organism exists to procreate; this is a prime directive. Once having achieved this, once having perpetuated the race and its own DNA code, the parent is on the downslide. It may still have child-rearing duties, but basically, the body is nothing more than a food packet that hosts many organisms while it's alive and feeds many more once it's dead. Human awareness of the fact that we are born dying usually results in an unconscious resentment toward parents (and toward a God figure), for although they gave you life, they have also given you certain death.

But there is one parent that doesn't have to worry about such resentment. When the amoeba—that single-celled, squishy-looking thing they were always talking about in biology class—"gives birth," its offspring is ensured virtual immortality. The amoeba, as well as others in its class, do not reproduce sexually. It divides. An individual's middle shrinks to an hourglass shape. Internal parts (food vacuoles, nuclei, etc.) start to split, with the emancipated parts sliding into the other end of the hourglass like two ends of taffy. Then the cell walls close together and sever the two halves. Now there are two individual organisms instead of one. Amazingly, both halves are younger than the original. Which in effect means that the original amoeba never dies—it is immortal. When scientists placed them on maintenance diets, they grew old and died; but if given an unlimited supply of tetrahymena (a simple plant), the amoeba theoretically lives forever. Every amoeba currently living—though it may have divided an hour ago—is millions of years old.

➾**FYI:** Longevity is simply a matter of growth: as long as an organism continues to grow, it lives. Once it stops growing, it descends into death. Giant **tortoises** often live to be three hundred years old, and for all but the last few years, they continue to grow. The same is true for the giant sequoia trees, which can live to be over four thousand years old.

ANTELOPE

People love to smell babies. They smell so . . . babylike. Some indescribable combination of freshness and sweetness. Perhaps that scent is designed to be compellingly pleasant, to make us want to continue smelling the child so that we can learn to distinguish our own child's smell from that of others.

The first thing the mother antelope does is learn her offspring's special scent. That is how she knows which of the young to nurse, for she will only nurse her own and no others. If her offspring dies, she holds a vigil beside the body until it decays to the point that she can no longer smell its distinctive scent. This is not unlike people smelling the clothes of loved ones who have died.

BABOONS

One of the most devastating moments for many parents is seeing their child with an open suitcase on the bed, packing up to head off for college or a job or just to begin their own lives. They are leaving home—for good. This behavior, called transfer pattern, takes place among all social primates. The reason for it is clear enough—if the young don't leave the group, then they end up mating with each other and this incestuousness is destructive to the survival of the group. Of course, it's not required that all the young leave, just those of one sex or the other. One sex goes off and joins a new troop, while the sex that stays behind mates with the newcomers from other troops. Hence, stronger offspring, stronger group. With **chimpanzees** and **gorillas**, the females usually leave the group to join another one while the males remain at home in the company of their parents. However, among the Old World monkeys (baboons, **macaques**, and **langurs**), the male leaves. Primatologists are not certain why it's different with different species.

The method of transfer is very painful for the one doing the transferring. At first, a troop of baboons may happen upon another troop, which is followed by a lot of posturing and threatening gestures. After that, both sides relax and ignore each other. A young adolescent male from one troop may linger a bit to watch the other troop. The next time they meet, he may sit at the edge of the group, and so forth until one day he just stays with them. As with most groups, there is an initiation process by which a group tests the newcomer. In this case, the test is for the male to be ignored, maybe even cuffed around a bit. But there is no intimacy, no friendship, and the newcomer is usually covered with parasites because he no longer has anyone to groom him. And since he is not a welcome sexual partner, he spends a lot of time masturbating. Eventually, though, he will no longer be hit; then he will be greeted by others, and eventually he will be a member of the group, with all its rights and privileges. On occasion, a brash young male will come in and violently bully everyone immediately, thereby gaining instant status in the group.

➾**FYI**: Among the **Bonnet** and **rhesus macaques**, the quality of parental care is dependent upon the sex of the child. Because adults are more aggressive toward mothers with daughters than toward mothers with sons, mothers tend to be more protective with their daughters. Among the **pigtail macaque**, adults are more aggressive toward females pregnant with female fetuses than those pregnant with male fetuses. It is not known why.

BATS

The old saying "It takes a village to raise a child" is appropriate here. After giving birth and nursing her young, it is unlikely a mother bat will ever see her offspring again once she flies out of the cave to forage for food. When bats return to their cave at sundown, they alight randomly. Females then nurse whichever infants stumble upon them first.

BEETLES

The **dysticid beetle** is one kickbutt predator. Like a nasty gunslinger in a lawless mining town, its ferociousness as a hunter allows it to dominate its home pond. It pretty much tries to eat every other insect it runs into. Its mating ritual is equally as aggressive. The male attracts a female by dragging his leg and wing across his own scaly body like a monstrous violinist. Lured by his song, the female approaches him; instantly he uses his suction-cup-like feet to attach himself to her. For the next few weeks, they remain stuck together like a cloying newly engaged couple. They do everything in this attached state: they hunt, eat, swim, and copulate as one being. When she finally lays her eggs, he releases her. The story begins to resemble *King Lear*: The biggest threat to the parents now is their offspring. The larvae are bigger than the parents and will devour their own parents if they get close enough.

BLUEBIRDS

For many men, one of the attractions of having a spouse and children is the notion of a happy, loving home to come back to every night after a long day's labor. Kick off the shoes, chat with the kids, snuggle with the wife. That's what makes the rest of the day's stress worthwhile. But for fathers among the **blue-backed fairy bluebirds**, there's no such reward. Because his plumage is so brilliantly colorful (a cowl of bright blue), he attracts predators. It's a little like living in a crime-prone neighborhood, with Dad driving a flashy Mercedes and wearing a gaudy gold necklace and a Rolex watch. It would be just a matter of time before some local toughs pay him—and perhaps the whole family—a visit. That's why the father blue-backed fairy bluebird doesn't go any closer than sixty feet to his nest, even though the nest itself is camouflaged with moss and lichen. Instead, he stands guard nearby to keep an eye out for predators. If one comes near he may streak off to divert its attention. Meanwhile, back home the female sits on the eggs. Because her job is time intensive, the male must also feed her. However, she must fly from the nest to receive the food in order to keep the nest's location a secret.

BULLHEADS

Who makes the better parent, the mother or father? With this fish, the father is convinced that he is the superior parent—and he's ready to back his claim up. The road to parenthood begins with the male digging a nest on the bottom of a stream under a rock. There he lurks, waiting for a female fish fat with eggs. To entice her, he bites her on the tail, which she finds sufficient enough foreplay to enter his little cave. Here she waits, often lying on her back for hours; when she finally lays her eggs, she does so on the ceiling of the cave. Now that she has done her job, the male boots the female from the cave. During the following two weeks, he fans his fins to aerate the water over the eggs, thus providing them with more oxygen. After they hatch, he alone cares for his offspring. (See also CATFISH below.)

CATFISH

The male **gaff-topsail catfish** takes parenting matters into his own—mouth. Found along the continental shelf from Panama to Cape Cod, this fish is distinguished by a large sail-like dorsal fin and a do-it-myself attitude about raising his offspring. After he has induced a female to lay her eggs, he then fertilizes them and scoops them all into his mouth. This can be quite tricky for a fish that is only twenty-two-inches long himself, for each egg is about three-quarters of an inch in diameter, and he may have to fit as many as fifty of them in his mouth. He carries these eggs for sixty-five days without being able to eat anything.

After they hatch, he still remains the ever-vigilant parent. He swims to the seaweed or sponge beds and spits out his young so they can swim around and feed while he guards them from predators. Sometimes a young one will become covered in silt, which can kill it by clogging its pores. Dad sucks his offspring into his mouth, rinses it off, and spits it right back amidst the others. After feeding time is over, he scoops them all up again and lies on the ocean floor while they all sleep.

The father is also able to distinguish his young from other fish the same size. An experiment was conducted in which a striped bass fry was introduced among the siblings. The father collected it along with his own. However, the next time the father released his young to feed, the bass did not emerge.

This amazing parental devotion continues until the offspring are about three inches long, at which point he spits them out never to return. That size is beyond the father's ability to recognize them as his own, so if he sees them again, he will eat them.

CHIMPANZEES

Even without children, maintaining a harmonious sex life with one's mate can be tricky; certainly the enormity of the literature about keeping sex lively in a relationship attests to that. Throw a child into the formula and the problem only gets worse. Humans often complain that the addition of a baby causes a strain on the relationship, not just from sleep deprivation or financial considerations, but because there seems to be diminished sexual desire and opportunity. For humans, part of the reason is that the infant requires so much attention. Chimpanzees have the same problem. Raising a baby chimp requires seven years of devotion, and because there are so many predators and food is often scarce, the mother is only able to concentrate on one offspring at a time. That means she can have sex only once every seven years. This may be one reason that chimpanzees are not a monogamous species.

CUCKOOS

Cuckoo birds are either the laziest or the smartest parents in the world. Rather than raise their own offspring, they skulk around waiting for the mother bird of other species (such as a **warbler**) to leave their nest and their eggs unguarded. The female cuckoo then flies into the other mother's nest, steals one of the eggs and replaces it with her own egg; this leaves the same number of eggs in the nest as before. The mother makes certain that the bird she invades has eggs of similar size and color. The returning mother doesn't notice the difference, that one of her eggs is an impostor. Nor does she seem to notice anything is wrong when the cuckoo chick hatches before her own; nor when that huge chick instantly begins to chuck the other still-unhatched eggs out of the nest, killing all the mother's original offspring. With the other eggs gone, the mother can now devote all its attention to the remaining chick, which she dotes on and feeds as if it were her own.

As for the cuckoo's birth mother, therein lies fodder for a dozen talk shows when the offspring grows up. The mother, free of any parenting burdens, flies off to have sex with pretty much any available male of her species. She could have a couple dozen such sexual encounters in a mating season, delivering a couple dozen eggs, each dropped in an unsuspecting foster parent's nest.

The **yellow-billed cuckoo** at least tries to be a good single mother. She actually builds a nest, but she's such a shoddy craftsbird that the nest usually crumbles apart in the first good rain. That's when she switches to Plan B: laying her eggs in another bird's nest. But unlike some of the other female cuckoos, she doesn't then fly away; she stays and helps the female whose nest she's invaded. Together they share duties sitting on the eggs until they hatch. Afterward, they share parenting duties, helping each other raise not just their own, but all the offspring.

➾**FYI 1:** When people think cuckoo—after they think about the clock—they think of a promiscuous bird with terrible parenting skills, as described above. Actually, of the 128 species of cuckoos, only 50 dump their eggs in the nests of other birds. The rest embrace the usual family values of most other birds, hatching and raising their own young.

➾**FYI 2:** The cuckoo is not the only bird that acts as a parasitic parent, exploiting others to raise its own offspring. The South American **black-headed duck** of Argentina and Chile also lays its eggs in the nests of others, usually ibis, herons, spoonbills, and other waterfowl. This little hatchling politely removes itself from the host nest shortly after breaking out of its shell, never to disturb the surrogate parents again. But the North American **brown-headed cowbird** leaves it eggs in a host's nest, which often leads to the complete destruction of the host's own offspring. The cowbird hatches before the host's eggs and therefore gets more food than the others, since it is able to scream louder. In its frenzy for food, the cowbird hatchling will trample the host's chicks to death.

FROGS

A tradition of high school biology classes used to be to have an aquarium with tadpoles so the students could witness the marvels of metamorphosis when what looked like a fish started to sprout legs and become a frog. But not all frogs develop from tadpoles. The **Andean Darwin's frog** begins life in the usual frog manner: The female lays eggs, and while clinging to her back, the male fertilizes them as they emerge from her. Now the female returns to the water, but the male gathers a group of other males and together they stand guard over the eggs for several days. Just before the eggs hatch into tadpoles, the males divvy the eggs, with each scooping some into his mouth and using his tongue to lodge them deep into his vocal pouch. The eggs stay safely inside these males for a few more days—nourished by their own egg yolk sac—until the 0.4 inch young crawl out of the males' mouths as fully-formed tiny frogs.

The female Australian frog, ***Rheobatrachus silus,*** also scoops up the tadpoles and develops them internally—not in her vocal sac, but in her stomach.

GIBBONS

Whether it is a culturally imposed stereotype or just a natural maternal instinct, the fact is that many girls play with dolls. We have dolls that cry, that pee, that laugh, that do pretty much everything a real baby will do, except dolls can be shut off whenever you want. All this doll handling is actually a form of basic training for females, a subtle course in how to handle their own babies. The gibbon ape doesn't have a doll, so they have a practice baby to serve the exact same function. The father of this practice baby is usually the female's own father (see also INCEST: GIBBONS). Through observation of her own mother with the other siblings, the young female learns how to treat the small offspring. However, mistakes are made and the baby gibbon usually dies. Yet, now she is prepared to better raise her own "true" baby, and it will have a much stronger chance of surviving.

GIRAFFES

Despite living in a peaceful, open society that demonstrates some of the human ideals of "a peaceable kingdom," the giraffe does not share human's interest in parenting. The fathers, since they are not monogamous, have no idea which offspring they have sired, so they have no proprietary investment. And the mothers are not much more devoted. It is rare to see the mother and offspring together. She raises them for a couple months, then pretty much leaves them alone. Should the young offspring decide to lope off and join another herd, the mother has no reaction. Though this seems callous from a human perspective, in a society in which there is a general acceptance of newcomers and very little violence, there is no need to regard one's own offspring as anything more than an extension of the herd.

➾**FYI**: As it was on the elementary school playground, so it is for giraffe social structures: the tallest is in charge. As the tallest giraffe strides among the herd, those who are shorter, and thus of a lower rank, must lower their heads and push them forward. If the lowly peon does not lower his or her head, there may be a fight, with each animal swinging his neck against the other's neck.

Other than that, the rules are pretty loose. If a giraffe wants to switch herds, he or she does so with no interference from Mr. Big. New members are welcome, without being made to feel like an intruder (unlike the closed societies of wolves and lions). Other than the head-lowering thing, there is very little aggression and fights occur rarely. The only stipulation in giraffe society is that each sex remains in a subherd of their own sex. No mixing.

HANUMAN LANGURS

The core of a pride of **lions** usually consists of a select group of sisters that is joined by one or more males, usually brothers (see also VIOLENCE: TURKEYS for similar brotherhood arrangement). When the brothers come in and take over, one of their first acts of office is to kill the cubs that the females had by other males. This induces the females to come into heat faster and to pay more attention to the new fathers' offspring, which is necessary because cubs have such a high mortality rate. Also, the males can't waste any time: Their tenure as leaders of the pride averages only about twenty-two months before a younger stronger group of brothers ousts them. That means they have only a limited time to impregnate the females and have their own offspring old enough to be on their own before they are thrown out of office.

The male Hanuman langur is a monkey that behaves similarly to the lion. A troop consists of twenty to thirty monkeys, and is made up of females and their infants and one male leader who mates with all the females. However, this exalted position makes him the object of constant attack from other males, especially all-male troops. The average tenancy as troop leader is only twenty-one months. As with the lions, time is crucial: so many females, so little time. So, like the lion, a male who takes over a troop filled with lactating females (which will keep them from mating) must first kill the infants. This will bring the females into heat sooner and the process can begin anew.

However, females are not always too passive when it comes to the murder of their children. One method they employ to keep the male from killing their children is to band together as a group in defiance. This sisterhood is accomplished by making sure every female achieves some bond with the other's offspring. Mothers pass their young around to the other females, who play with them as surrogate mothers. Once this bond is established, the females are more likely to defend the infant against the murderous intentions of the male. In such a case the male will often back down. When this doesn't work, the mother may even risk leaving the safety of the troop in order to save her infant.

KANGAROOS

There's a lot of psychobabble about people subconsciously wanting to get back to the womb—a place of snug security and protection. Female kangaroos provide a nifty substitute: a furry pouch. The kangaroo is a marsupial (a mammal that carries its young in a pouch); she gives birth, not to a fully developed baby kangaroo, but to a slug-sized embryo. The mother may lick a path to her pouch for the embryo to follow (which takes it about half an hour of trekking), or she may just pick the little thing up in her lips and deposit it into her pouch. Either way, the offspring will immediately clamp onto one of the mother's four nipples. The nipple instantly swells to form a tighter attachment between mother and child. This is crucial for two reasons: (1) to protect the infant from the bouncing movements of the mother and (2) to discourage the males who often snatch the infants from their pouches and fling them away to die. If it avoids death, the offspring remains in the pouch for about eight months, growing to nearly the size of its mother. Even at that size, it will get out of the pouch, graze with the herd, then return to the pouch, its huge head protruding as mom hops away, sluggishly bearing her heavy load. (See also SEX/KINKY: KANGAROOS.)

➾**FYI:** If the female is carrying offspring in her pouch when she conceives again, the new embryo will develop until it reaches about one hundred cells (about the size of the tip of a ballpoint pen). At that point is stops—and waits. It will not grow again until the older offspring living in the pouch either dies or leaves the pouch. Since many baby kangaroos die within their first year (due to droughts, hostile males, and human encroachment), the newly forming embryo usually doesn't have to wait too long.

LIONS

Tarzan movies may make you think of bushy-maned lions as stealthy hunters chasing scantily clad women or corrupt diamond hunters, but real lions don't like to put out that much energy. The bushy-maned ones are the males, and they generally let the females do the hunting, content to let the lionesses drag home dinner while they wade in, push the females aside, and are the first to dine. Although this behavior may seem a little insensitive, there is a good reason for it—the survival of the species.

Lions live in small groups called prides that consist of one to four males, several females, and their cubs. It is the lionesses that hunt, give birth, and raise the cubs. The male does not involve himself much with these duties. He does not hunt much because his huge mane might attract attention and frighten away prey. (The burden of good looks occurs likewise with some colorful birds that stay away from their nests and young because their gaudy plumage attracts predators.) Therefore, when the pride is out stalking food, the lionesses are out in front, followed by the cubs, followed by the males. This way the males can protect the cubs, who generally lag behind their fast-moving mothers.

But the male lion's job as a parent is still not done. Once the carcass of the gazelle is ready to eat, the male pushes in and claims a large portion, which he will share with the cubs. Lionesses are not especially generous mothers; they may bat their own cubs away from the food while they continue to eat. In this case it is the father who keeps the cubs from starving to death.

➾**FYI:** Although the father is conscientious about providing food for the cubs, he is just as ruthless about denying them food if game is scarce. During times when there is a short supply of food, such as a drought or unsuccessful hunting, the pride will let the cubs starve to death. In fact, the main cause of cub death, even in normal times, is neglect.

MIDGES

"How sharper than a serpent's tooth it is to have a thankless child," says King Lear, speaking for parents through the ages. For children this statement can be viewed as another guilt-inducing complaint from the folks. But among midges, this comment has a frightening reality. As soon as the female midge becomes pregnant, her brood begins to digest her from the inside, eating her internal organs until she is nothing but a husk, like a piñata, from which the well-fed offspring will burst forth, leaving behind the tattered debris that once was their mother. (Before you shed tears for poor Mom, take a look at her mating habits under SEX/DANGEROUS.)

OPOSSUMS

We know that courtship is often a competition through which the best mating material is determined. Those who do well get to have sex and pass along their genes; those who don't do well never have sex and their inferior genes die with them. This same competition can be extended even among newborns (or even embryos—see VIOLENCE: SHARKS).

The female opossum can give birth to as many as eighteen eraser-sized babies. The problem is that mom has only enough room on her nipples for seven. That's why these little creatures must hustle across her pouch (opossums are marsupials) as fast as they can. Only the first seven get to live; the others starve to death.

PENGUINS

Some species of male penguins easily get the Mr. Mom Award for most dedicated dad. The female penguin lays an egg that must be incubated between the top of the feet and the warmth of the belly. However, once she's laid the egg, the female doesn't stick around for the tedious job of balancing it on her feet; she turns those duties over to the male. As soon as she does, she beats a hasty retreat for the sea, which can be as much as fifty miles away. Once in the water she will feed for two months, fattening herself with undigested fish with which to feed her offspring upon her return. Once she returns, the male relinquishes his vigil and heads straight for the water and his first meal in months.

➾**FYI**: Like the penguin, the male **West African black-chinned mouth brooder** is a dedicated father who sacrifices comfort to raise his offspring. After the female fish has laid the eggs and the male has fertilized them, the male scoops the eggs into his mouth and carries them there until they hatch twenty-three days later. During this time, he is unable to eat.

SEALS

Shakespeare's *King Lear*—quoted above—is considered his most devastating tragedy. It is the story of a king who mistakes power for wisdom and in his effort to defend this power destroys everything he loves. It is only when he loses his power that he gains the wisdom he thought he had. Unfortunately, he causes the death of his loved ones first. The lesson is universal: the more things we gain—be they power, money, possessions—the more we have to work to defend what we have. That can lead to tragedy—even in the world of seals.

The male **elephant seal** is an enormous creature by necessity. First, because during mating season he stops eating altogether and lives off his stored fat. Second, because he must constantly fight the other males—sometimes to the death—in order to establish dominance. This is especially crucial since only the dominant males get to breed. To the victors go the harems of females, and the rest of the males slink off to watch and wait for the time when they will dominate. However, the very qualities of size and aggression that made this dominant male so attractive to the females also makes him less than an ideal parent or spouse. In order to defend his territory, the hulking mass will sometimes crush his own children and spouses as he lumbers over their bodies to attack an invading male. Almost half the seal pup deaths each season result from the adult males crushing the pups to death.

SQUIRRELS

One of the psychological barriers men exhibit in their reluctance to have children is the fear that the offspring will replace them in the affections of their mate. In other words, the woman will be so busy with and enamored of the child, she won't have the time, energy, or love for the man.

Female squirrels have the reputation for being among the most loving and affectionate mothers in the animal world. However, they are not very loving toward males. So, the male strategy for evoking affection from the female is to woo her by behaving like a baby squirrel. He starts this behavior right before mating and continues it for several days afterwards. It works, too, because the female immediately shows him the same devotion she would show her offspring.

To put this squirrel behavior in context, it is important to know something about their mating habits. The female sends out a call that she is ready to mate and as many as a dozen males start running after her. This is not unlike a woman wearing a microskirt and plunging neckline flirting among the male patrons of a sports bar; she will have sent out a signal and many of the men will respond. And, as can happen among men in a bar, the male squirrels will fight each other as they pursue the female. They chase her up and down trees and through the woods. There is no danger of rape here, for the female's bushy tail covers her vagina like a vault door; she must deliberately move it aside for there to be sex. This chasing scenario is designed to eliminate the weaker suitors until the female is standing face-to-face with the victor. But once the most macho of the squirrels emerges, he turns to pudding. He begins acting like a baby squirrel, which serves as a catalyst that switches on the female's maternal instinct so she can overcome her aggression/aversion instinct. Only then can they mate. Afterwards, they stay together for a few days, the length of time it takes for her aggression instinct to recharge itself. At this point she chases him away.

➾**FYI:** When the male **jackdaw** bird angers his mate, he wins his way back into her good graces by imitating a needy baby jackdaw. This adopting infant behavior to win affections is fairly common behavior among male, and sometimes female, animals. This explains the penchant among courting humans for talking baby talk to each other; they are in fact biologically hardwired to act like children when they are in love.

WASPS

The female **jewel wasp** goes through a lot of trouble to provide all the necessities for its offspring, including an incubator, a condo, and plenty of food. It all begins when the pregnant wasp attacks a **cockroach** (or other appropriate insect), stinging it into paralysis. Then, while the insect can't move, she snips off the tips of its antennae, disorienting it. She skitters off to her small cavern, perhaps under a rock, and clears away the tiny pebbles (big as boulders to her) that guard the entrance. She hurries back to the still-stunned cockroach and guides it into the cavern. Now she lays her single egg on the cockroach's body, which is still too poisoned to do anything about it. The mother then leaves the cavern; once outside, she replaces the boulders around the entrance, essentially entombing her child with the cockroach.

But the larva is not defenseless. In fact, being walled in protects it from other predators and allows it time to grow. As it grows it feeds off the cockroach, boring through the insect's body to snack on its insides. Once the wasp is grown, it emerges from the empty cockroach husk strong enough to dig its way out of the cavern. Mother and child never see each other; at least they don't recognize each other as such.

ZEBRA FINCHES

Male zebra finches can't recognize females. This can have disastrous results come mating time. Male **guppies** have the same problem, which leads them to sometimes accidentally court predators of other species, resulting in entire male populations being wiped out (see also SEX CHANGES: GUPPIES). But male zebra finches have the advantage of some tough-love parental training. This training doesn't occur until after the young male finch is over thirty-five days old; if he's removed from his parents before then, he will never learn how to tell the females from the males. If that happens, he might as well buy a satellite dish with two hundred channels because he's not going to be dating much. However, if he is still at home, somewhere between thirty-five and thirty-eight days of age, the father starts to get nervous. Before this time Dad was an extremely attentive parent, but now he begins to look on his own son as a potential sexual rival. Get out your Oedipal handbooks, folks, because this is where Freudians go to town: The father begins to peck at his son in a constant onslaught until he finally chases the offspring from the nest. Forever afterwards, in self-defense, the son fears any bird that looks anything like Dad. Since only the males look like Dad, he knows to stay away from them and cozy up to the other ones—the females.

10
PROSTITUTION

Field notes. . .

There are plenty who would define traditional marriage as legalized prostitution: women exchanging sexual favors in order to be supported by men. Even courtship is sometimes a display of economic bartering: The male picks up the tab and often thinks he's entitled to something in exchange, generally some physical act of affection. In some cases, nature finds this activity quite acceptable, for in providing the female with some token, the male is proving his ability as a provider. This enables each "lover" to assess the other's potential as a mate and parent. For example, some birds, like the **common terns** and **sandwich terns**, go through a lengthy courtship before eventually pairing off for a monogamous "marriage." During this courtship, the male tern catches fish and gives them to his mate, thereby proving he will be able to feed her and the chicks later.

Strictly speaking, the term prostitution may be unfair here. It implies a moral judgment on what is a natural instinct on the part of many species. At what point is it common courtship behavior and what point does it become prostitution? For humans the line might be drawn when a woman performs some act (i.e., sex) that she wouldn't have done without a gift—the gift therefore being the only goal of the courtship. If the gift is the main motivation for bestowing affection, that's prostitution. Most of the examples given here are demonstrations of "courtship feeding," a ritualistic exchange of food during mating that further cements the bonds between a couple. Some biologists even suggest that this ritual is the origin of human kissing.

Instances of courtship feeding that are more ceremonial are included in the COURTSHIP chapter. Those that are more directly cause-and-effect have been included here. Since other animals make no distinction between courtship and prostitution, this chapter is purely for comparison with human notions.

BEE-EATER BIRDS

A current ad for diamonds encourages "Give her something that says you'd marry her all over again." Obviously from the pronoun, this ad is directed toward men, and though the sentiment may be heartwarming to some, it subtly echoes one of the consistencies in nature: gift giving in exchange for sexual favors. A box of chocolates or an expensive dinner don't entitle a man to anything, but it is obvious that there are subtle expectations attached to this courting process. This is even true when the couple is married: How many comedies have we seen in which a rich, wimpy husband is refused entrance to his wife's bed unless he first brings her expensive jewelry?

But the **European bee-eater** is a little different than other examples of prostitution. The male and female have already become mates, that is they have agreed they are a couple, though they still haven't had sex. After they become virginal mates, the couple spend the next couple weeks in hard labor, building a nest. Now the honeymoon may begin—but not until the male comes up with a suitable gift. Only after the male has presented his betrothed with a dead bee whose poison gland has been squeezed dry will she allow him to mount her. Thus the male proves he can build a nest and catch bees.

His ability to catch bees is crucial since it will shortly be his job to feed his spouse. Once the female lays her eggs, the male and female alternate sitting on the eggs. When the female is off duty, she has the time and skill to catch her own bees, but she prefers not to. She waits until she climbs back onto the eggs to eat, for that is when the male—who has just finished his tour of duty—must now spend his break time hunting bees to feed her. After she's eaten the bees, the female hacks up the bee's armor, which she then uses to line the nest.

➾**FYI:** The **bonobos monkeys** also engage in this form of sex for barter. Food is given to the female, sex ensues.

FLIES

The male **hanging fly** presents captured flies to the female as a preamble to sex, not unlike humans with a box of chocolates. Once he's snagged the fly, he zips through the air triumphantly, releasing a scent to which the females are attracted. As the female comes toward him, he hauls out his gift and presents it to her. But she's picky, accepting only the largest flies. Only when she has begun noshing on the fly is the male able to mate with her.

The male **dance fly** (so called because females gather in groups that move in a sort of frenetic dance) must also bring a gift if he expects to get lucky. He may bring her an insect or just part of an insect. As she feeds on the gift, he hurriedly mates with her, since he must finish copulating before she finishes eating. Otherwise, he may be dessert. Other species will bring an elaborately wrapped package that resembles a silken balloon (the silk having been spun from the male's anus). The extra time it takes for the female to unwrap this gift allows the male extra time for sex. The more roguish of males don't even bother to include an insect in their package; when the female finishes unwrapping, she finds nothing inside. By then, though, the male has finished copulating and is far away. (See also SPIDERS below.)

➾**FYI**: In yet another example of rituals that survive apparent practicality (see also COURTSHIP: CUCKOOS), in some species of dance flies that are vegetarian, the female will eat meat one time in her life—when her mate presents her an insect as a wedding gift. There is no useful reason to bring such a gift since the male is in no danger of being eaten by her; he requires no distraction to save his life. Also, as gifts go, this one stinks. She doesn't even eat insects! Still, as Tevya in *Fiddler on the Roof* might explain with a shrug, "It's tradition." Some ancient knowledge remains from the time when members of this species were carnivores. And, yes, she eats it—the first and last meat of her life.

HUMMINGBIRDS

A 1969 study of the **purple-throated carib hummingbirds** of the West Indies revealed that during the non-mating season of winter, the males guard the fields of flowers that provide the best nectar. The females will then approach the males, inviting sex. However, before the actual sex can begin, the female is allowed to feed on the nectar from the plants. She will return again and again in this sex-for-food exchange, even though she has not yet produced her eggs and cannot possibly conceive.

PHEASANTS

Arranged marriages have often been promoted on the philosophy that, though a couple may not be in love at the time of their marriage, they will grow to love each other over time. The male pheasant has a similar attitude, for it only begins to court the females *after* it has mated with them. On the other hand, because their "marriage" only lasts a short time, the male is on a pay-as-you-play plan with his wives.

The cock has a harem of two or three hens that he walks with daily during spring mating season; they always take the same route. Every once in a while, the male will stop in front of a bit of food and offer it to one of his wives. If she takes it, they mate. Two weeks later the hens start to lay their eggs and the harem is no more. The male, with his bright plumage that attracts predators, is no longer welcome.

ROADRUNNERS

This chapter is filled with examples of females that refuse to have sex unless first presented with a gift, usually of food. However, the roadrunner is unusual in that the male withholds payment until after he's had his way with the female. The male roadrunner races through the deserts of California, Arizona, and New Mexico, snatching up mice or baby rats. When he catches one, he smacks its head against the ground until it is dead or dazed. Then he dangles it in front of the female he's interested in. Sometimes she'll beg for the food, but he is not so easily persuaded. He keeps the food away from her grasp while he copulates with her. Once they've finished, he feeds her the prey, which now will help nourish the eggs he's just finished fertilizing.

SPARROWS

As we have seen with many birds, cool abodes talk and chirping walks. In other words, those males with the nice nests get the females and those without do without. It's not just a matter of a nice nest—it's a matter of having any nest: Not all male sparrows have nests and without a nest they have no chance at all of ever mating. When a landowner sparrow sees a female he's interested in, he chirps to get her attention then steps aside so she can inspect his abode. She does this without paying any attention to the male; after all, his character as a potential mate will be measured by his handiwork, not by his appearance or song. Very often the nest will not meet her demanding standards and she will fly. (See also COURTSHIP: BOWERBIRDS.)

SPIDERS

The female of some species of spiders is much larger than her male counterpart. As a rule, the bigger the female compared to the male, the greater the risk of her eating him before, during, or after sex (see SEX/DANGEROUS: SPIDERS). Among some species, the males have devised a scheme that is not unlike paying tribute to a god, or merely paying off the neighborhood bully. The male ***Pisauridae,*** for example, catches an insect, wraps it up in silk as if it were a Christmas present, and gives it to the female. While she's busily occupied unwrapping her gift, he charges in, has sex, and takes off. Sometimes, if she's a particularly slow unwrapping, or he's especially quick having sex, he grabs the partially unwrapped gift as he runs away—having just managed a freebie. (See also FLIES above.)

WOODPECKERS

As with the sparrow, the female **greater spotted woodpecker** judges the male as mating material by his ability to build a suitable nest. The staccato hammering sound that we're so familiar with as characterizing the woodpecker (such as at the beginning of every Woody Woodpecker cartoon) is the male's tribal announcement over his P.A. system indicating he's finished his nest and that any males who come near it are going to receive a pecking like the one he's delivering to the hollow tree. Generally, other males will stay away since they don't usually fight. If they do fight, however, it is often to the death, with each hammering the same staccato blows to the head of his rival as he would to a tree.

Courtship can take several months since each partner must overcome its fear of the other's aggression. The male's pecking scares the female and she usually flies away; he then chases her. Eventually, she will land on the same tree where he keeps his nest, he will perform a ritual courtship flight, and then she will inspect his nest. A woodpecker without a nest has no chance of mating.

Once they do mate, they still don't like each other very much and keep their distance to avoid sparking their mutual natural aggression. However, both are conscientious parents.

➾**FYI:** The **American red-bellied woodpecker** has a slight variation on this mating ritual. When the female flies to the male's tree, the male ducks inside so she can't see him. Now they each peck on the wood from opposite sides of the trunk, like teenagers chatting on the telephone. And like teens, they find it easier to communicate when they can't see each other than when they can. Once they mate, they rarely inhabit the nest at the same time.

11
RAPE

Field notes. . .

For much of human history, rape was considered such an abominable crime that it was punishable by death. The methods of execution could be pretty grim, often involving mutilation of the offending body parts before or after death. Though the United States stopped executing for rape in 1977 (having put to death 455 men for rape since 1930), several countries around the world (El Salvador, North Korea, Malaysia, South Africa, etc.) still treat it as a capital offense. In some countries where rape is not technically punishable by death, judicial systems look the other way when a family member avenges a rape by murdering the alleged rapist. Ever since we've formed societies, that's the way it's been pretty much everywhere. Whether or not they employ the ultimate punishment of death, most human societies have strict laws about what constitutes rape—either when it involves forcible sexual activity on an unwilling person or even consensual sexual activity involving a minor. In general, humans are in agreement: Rape is wrong!

But before we congratulate each other on our righteous attitude in protecting hearth and home, we might want to examine the issue a little closer. First, we haven't always been, nor are we yet, in agreement as to what constitutes rape. Laws vary, with some countries arguing that if a husband has sex with his wife, even if it is against her will, that cannot be rape. This was the law in the United States until recently. Also, there is still the problem of public opinion; women who dress provocatively or who go to a man's home are often thought to be "asking for it." One recent survey indicated that 86 percent of American boys age thirteen to fifteen think it's okay for a husband to rape his wife; 24 percent say it's okay for a man to rape his date if he has spent "a lot of money" on her.

Also, we need to examine why we have such harsh laws regarding rape. The original motivation for these laws against rape had little to do with protecting individual integrity and a lot to do with protecting genetic lineage. Men did not want any other males imposing their sperm—and therefore potential genetic code—in either their wives or daughters. Obviously the wife's fidelity was sacred because sex with any other male than the husband confused who fathered the offspring; this could create havoc with inheritance, not to mention the hard-wired desire to pass on one's own genetic code. This is why rape was punishable by death in most religions and in most societies (see also ADULTERY). The rape of one's daughter was also a major crime because it damaged, if not destroyed, the marriage value of the daughter. Virginity was a key marketing attribute when marrying off one's daughter because the groom could be sure his would be the only sperm to enter the womb and therefore the offspring would definitely be his. Marriage in most parts of the world, both historically and currently, is as much between families as it is between the bride and groom. The bonds of marriage between royalty united kingdoms, and these same bonds united families who were now bound to help each other in business, battle, and all else.

Even defining rape is argumentative. Some people are still debating whether forcing one's wife to have sex against her will constitutes rape or wifely duty. Certainly if a wife or husband refuses to have sex with their spouse, this is grounds for divorce in most countries. This attitude is also true of most religions. Sex is necessary, the thinking goes, in order to have offspring, which is what God (or the natural order) expects of us. It is also necessary for some in order to express passion and affection, thereby creating a more stable husband-wife relationship. However, religions do not condone forcing a spouse to have sex; they merely allow for divorce or annulment of a marriage if one partner refuses to have sex or children with the other (or, in some cases, if the spouse cannot conceive children).

It is commonly noted by experts that rape among humans isn't about sex; it's about violence and control. Certainly among many animal species—including humans—males will sometimes rape other males to show dominance. But it is more likely that among animals other than humans rape is about sex, because sex is about procreation and passing along one's genes. Mating sometimes includes actions that we would define as rape, but which are a common behavior pattern in various species. In fact, they know no other way; aggressive stimulation (i.e., biting the neck, punching, etc.) is required to facilitate the sex act, even to make it possible. Scientists have recently discovered that rape may have been the normal method of sex among many prehistoric dinosaurs. On occasion, such violent sex may have resulted in the death of one of the sex partners (see COURTSHIP for more). Scientists theorize that this aggressive behavior occurred because dinosaurs had not yet developed the ritualized courtship behavior that allows animals to overcome their aggression toward their own kind. Whatever the reason, such behavior still exists today. However, in the animal world, scientists avoid the term rape because it includes moral judgment; they prefer terms like "resisted mating," "forced insemination," and "willing resistance."

DUCKS

During spring mating season, male ducks become nearly uncontrollably lustful, pursuing females with unrelenting vigor. Even though ducks form permanent lifelong marriages, males still pursue other females during this time. In fact, the male is too busy chasing other females to sit around home protecting his own wife from pursuers. Usually the female is faster and more agile than the male and can therefore elude his advances—unless she welcomes them. Sometimes a group of males will attack a female, chasing her until she is so exhausted they can each have sex with her. Sometimes the female will kill herself in her frenzied attempt to escape from them. A **blue-winged teal** was once seen diving into a lake to avoid her male pursuers, and she was not seen to emerge.

➾**FYI:** Ducks are one of the few groups of birds in which the male possesses a penis.

GUPPIES

Yes, these are those cute one-and-a-third-inch-long fish that look like tadpoles and are often raised in schoolrooms as class projects. Technically, what the males do is more sexual harassment than rape, but it is of such an intensity that it edges close to rape. It starts with a fundamental difference between the genders regarding frequency of wanting sex: females are rarely interested and males spend fully half their time attempting to mate. If TV talk shows are any indicator, this seems to be a problem humans often face, though it is not clear whether among humans the problem is biological, cultural, or even mythological. Among guppies, however, it is quite real. The female is only interested in sex if she is a virgin or if she has just given birth. The rest of the time she forages for algae, eating as much as she can because the number of offspring she gives birth to will depend on how well she has eaten. But with males constantly harassing the females by trying to mate with them, the females spend a lot of their time trying to avoid males—females are approached sexually about once every minute! This constant barrage of unwanted attention cuts down their eating time by one-fourth. It's a little like a woman going to a restaurant to eat, but every minute she's there, a different male tries to mate with her.

Scientists speculate that the males who do most of the harassing are the pocket-protector-type nerds of guppyland who are unable to find willing females to mate with. Females are choosy about their partners, which means a lot of the males are excluded from mating. So these losers continually harass the females hoping that eventually the odds are they'll make a sale. As for the females, they are larger than the males and could put up a formidable defense by biting back at the males. But they don't—they prefer to simply swim away.

HARES

Although we tend to portray hares as fluffy, timid, nose-twitching cuties, they engage in a rather brutal form of courtship. In general, they are not a gregarious animal, often choosing a solitary life away from others. Except during mating season. Then they are exceptionally promiscuous. The female in heat runs away from the male, who pursues her until he corners her. Then he proceeds to beat her into submission, finally mating. Afterwards, they separate.

LOBSTERS

Lobster sex generally involves the male raping the female. The tricky part is that the male has no penis and the female has no vagina. At first, the male approaches the female by lightly stroking her with his antennae. If she doesn't run off, he becomes more aggressive, grabbing her with his claws. She struggles to escape while he tries to force her over onto her back. In the end, the larger, stronger male dominates and he proceeds to pin her down by clutching her claws in his and securing any other wriggling appendages with his four pairs of walking legs. He presses his abdomen against her tail to steady her, then penetrates her with his first pair of walking legs. The legs fit easily into corresponding pocketlike openings between her walking legs. His legs have canals allowing for the flow of sperm from his testes located under his shell. Once he has penetrated her, he may keep her pinned down for as long as an hour until he has completely filled her pockets with his sperm. After he releases her, she flips herself back over and scurries away.

➾**FYI**: Lobster breeding season is in the fall, but a female lobster will carry her rapist's sperm in her throughout the winter, not using it until the following spring.

MOSQUITOES

Prepare yourself. By human standards, this may be the most monstrous form of rape imaginable. In one species the male mosquito hunts for female pupae (the cocoon stage before the young come out). When he finds one, he lands on it, uses his dagger-sharp penis to slice open the cocoon, and then begins to have sex with the still-developing infants.

OCTOPUSES

The octopus has become a standard symbol of the groping adolescent male pawing at his date's body with a seeming pinwheel of arms. As it turns out, the symbol may actually fit the animal itself. The male octopus has no penis. Instead, one of his eight tentacles is constructed a little differently from the others and serves as the copulatory organ; this particular tentacle is easy to spot because it's longer than the others. Inside his body are little packets of sperm that, once placed inside the female, will rupture and free the sperm. Often the male will approach the female and gently caress her with his arms; both of them will become excited, and their bodies will continually change color. This is when the male slips his reproductive arm into the female's siphon (the breathing tube just under her head). Even though she will be breathing hard, she may allow this and all will proceed smoothly. The packets of sperm will travel from the storage area in the male's huge head, through the groove in his arm, and into the mantle cavity of the female.

However, sometimes the male simply leaps on the female, pinning her down with the webs between his arms. She fights back, each of them choking and tearing at the other with their tentacles or pecking at each other with sharp parrotlike beaks. Sometimes they fight until one dies; sometimes the wrestling is so violent that the male's reproductive arm is torn from his body and the female swims away with it still inside her. Eventually, the arm will be absorbed into her body.

It is sometimes said of human males that their penis is like a separate being. With the male octopus this is sometimes literally true. The reproductive arm will load up with a supply of sperm packets like a child packing up sandwiches to run away from home—and then it does just that. It breaks away from the body and swims away as an "independent" creature. When it finds a female, it enters the siphon, deposits the payload of sperm, and then dies, hanging there. This was the reason for the sightings of the mysterious "nine-armed octopuses."

SEALS

Humans have seen plenty of courtroom defenses in the past decade that involve battered-wife syndrome. This occurs when a woman has endured years of systematic terrorism at the hands of her husband, including repeated beatings and even rape. Under such circumstances, women who have struck back by killing or mutilating their husbands have been found not guilty due to the diminished capacity they suffered from such abuse.

The male **northern elephant seal** weighs eight thousand pounds, four times as much as the female. That is the equivalent of a 140-pound female making love to a 560-pound male. Because of his sheer size and power, the elephant seal has never had to develop much finesse in the wooing department. Whatever he wants, he takes. Which is how the bull seal manages to gather a harem of some forty cows. However, the females are often not all that happy with their circumstances and try to leave the harem. When one does, the bull chases her down, biting her and pummeling her with his flippers until she returns. Sex for the bull is just as direct. When he's horny, he grabs a female, wraps her up in his flippers and climbs on. He's usually finished within five minutes, but during that time the female often struggles to escape from him, kicking and screaming the whole time. However, there will be no "burning bed" or Bobbitt-izing here. The female simply endures. (See also PARENTING: SEALS.)

SKUNKS

Even consensual sex among skunks is a rough business. The male begins by stalking his victim, following her everywhere, sniffing and even licking at her genitals. (When the female is in heat, her vulva swells and secretes mucus.) Eventually she tires of his attention and attacks him. The two endure several skirmishes that do little damage. Weary of the foreplay, the male may then attack her, clamping onto the scruff of her neck. Then he mounts her, though he can't actually penetrate her without her cooperation (his penis is only an inch long and about as thick as the wire of a paper clip). Penetration is accomplished by scratching at her vulva with his rear leg until she is duly aroused and opens herself to him. So far, nothing unusual in all this. However, as the female moves out of heat, her interest in sex diminishes and she begins to fight the male off with increasing vigor. She rolls onto her back biting and kicking at him, sometimes inflicting painful damage to his neck and face. He may persist anyway, biting her neck and trying to roll her over for penetration. However, the male is rarely successful. Instead, the thrashing he receives only causes him to avoid her whenever they run into each other in the future.

TORTOISES

For most tortoise species, female cooperation is essential due to the delicate mounting procedure that involves the male's flat belly resting on the female's rounded shell back. For penetration to occur, the female must stick her rear end out of the shell as far as she can. With many types of tortoise, if the female is unwilling to have sex, she may press her legs together, swim ashore, or flip herself upright in the water, facing her suitor with her belly. These rejection techniques are generally successful—"no" means no.

However, during breeding season the male **land tortoise** will sometimes not be denied, whether or not the female is willing. If the female refuses his advances by tucking her butt into her shell, he walks around to the front and snaps at her head and bites her forelegs. In defense, she pulls her head into the shell. Ordinarily this defensive posture might be sufficient, but during breeding season the tortoises are plumper than usual and now she can't fit both ends into the shell. So, when she pulls her head in, her butt pops out and the male now goes back and mounts her. Sometimes she resists so much that the male continues biting her until she's lost so much blood that she dies.

➾**FYI**: Male **sea tortoises** can be especially horny during mating season, hot to copulate with anything in the water. On occasion, they have even tried to mount scuba divers.

12
SEX

Field notes. . .

Sex happens. But why—and why in so many variations? The physical body is hardwired to do one thing—have sex. Procreate. Everything else it does—eat, sleep, form societies, etc.—is nothing more than a support system to create an environment to procreate and raise offspring. Think of the planet—and perhaps even the universe—as nothing more than a giant corporation—like IBM or AT&T. Once a corporation is formed, it's sole purpose is survival; it must continue to exist. Every rule of the company, every person that's hired, everything is ultimately to ensure the continued survival of the corporation. Sometimes a particular branch may do poorly and have to be shut down—as when a species becomes extinct. When this happens, the corporation merely shifts its energy to other divisions.

So too the purpose of animals, on a purely physical level, is to produce more of their own kind. Why? For food. Each organism, whether bacteria or human, is nothing more than a packet of food that provides nutrition to hundreds of thousands of other organisms from the time of conception to well after death.

Each body walking, flying, slithering, or oozing around this planet feeds many others just by existing. You're sitting there reading this and thousands are living on your skin, in your intestines, between your toes. If you're in bed, the skin you're shedding is feeding the two million dust mites that are in bed with you right now. Your skin snows down on them and they devour it greedily. And so the great circle of life so celebrated in *The Lion King* is merely a cafeteria of mutual munching. Individuals come and go—species come and go—but the great corporation Earth, Inc. continues on without noticing.

Humans prefer to romanticize sex. That's because we are one of the most aggressive species on the planet, and in order to mate we require elaborate courtship rituals to break down that aggression level between a couple (see COURTSHIP and MARRIAGE). Another reason our courtship ritual is so exacting is because we take so long raising our young. First, we have to overcome our natural aggression toward the opposite sex, then we have to stay with that person to parent for a certain number of years. It would be terribly inefficient to have to go through the same elaborate process each time we wanted to mate and create offspring. If we did, it would cut down significantly on the number of offspring humans produce and thereby threaten the survival of the species. Romanticized or not, the type of sex a species practices is merely a matter of mathematics. It's the method that produces the most and healthiest offspring.

And there are many, many methods of sex. That is one reason this chapter is the largest in the book. Another reason is that much of human morality is directly related to matters of sexuality—some would say that's what all our morality is based on. Because sexual desire arises without our conscious control, it has come to represent the baser nature of humans. Much of our moral teaching involves lessons on how to control desire, rather than let it control us. Others argue that too much control of desire is unnatural and therefore breeds violence and depression.

We know that there is much more involved in sex than meets the eye. Despite all our love poetry and self-help books about how to select the best mate, there is evidence that we select mates based on their genetic makeup being the opposite of our own, which produces stronger offspring. We are able to recognize this genetic difference through smell, the one sense that is directly connected to our brains for instant access. The right smell—we can distinguish ten thousand scents—triggers an immediate physical response. And there's the psychological influence: Characteristics of an adult we admired when we were small children are unconsciously added to our requirement list in selecting a mate. Regardless of all the lip service we pay about having a good sense of humor, being sensitive, and loving children, we are really responding to stimuli over which we have no conscious control. If that weren't true, people probably wouldn't split up as much.

There are other natural factors involved in sex. What makes animals feel more sexual some days than others? One of the major factors is the weather. The length of the day triggers the mating drive in some animals; it lets them know it's spring or fall and time to mate. For the **crested bustard**, sexual desire is triggered by rain. This bird comes from the Kalahari desert of South Africa, where rain is infrequent. Therefore, if the bustard mates at the wrong time, its offspring will die in the drought. So it waits until it hears the sound of rain and then begins its courtship. In fact, in some zoos the bustard begins its courtship dance at the sound of nearby animals being hosed down.

What is "natural" sexual behavior? It seems like everybody feels qualified to answer that question. But below are subcategories involving sexual behavior among animals that challenge some of our cherished notions about what is "natural" in the world.

Types of Sex

1. *dangerous*
2. *duration & frequency*
3. *gender bender*
4. *group*
5. *kinky*
6. *positions*
7. *weird*

1. Dangerous Sex

You've seen the movies (Sharon Stone in Basic Instinct*) and you've heard all the jokes about leather, chains, and whips. You've seen news reports about snuff films, or about people having sex while choking each other to near unconsciousness (and sometimes accidental death). You ain't seen nothin', folks. The animal world provides a variety of sexual challenges in which the slightest misstep results in death. Talk about performance anxiety.*

Mythology mentions Cupid and his arrows of love. The metaphor describes how love feels: an exquisite pain of insatiable longing. Well, on a less metaphoric level, sex among some animals is literally a pain, sometimes even fatal, involving arrowlike penises that indiscriminately stab through the "love" partner's body (e.g., bedbugs and snails). It also involves some lethal sex in which the female devours, shreds, or castrates the male. Only the most sexually secure need proceed.

ANGLERFISH

If ever there was a species that might evoke all the castration fears in human males, the anglerfish is it. Never has all the "relationship" jargon of "I've got to have my own space" or "I've got to be my own self apart from you" had more dramatic meaning. The female anglerfish has a face more fearsome than any shark. With long, pointy teeth like spikes, some species (such as **Johnson's black angler**) are able to swallow prey twice their own length. This frightening appearance is all the more formidable when one realizes that the male of the species is only a tiny fraction of the female's size. While she may reach three to four feet in length, the male is a paltry half-inch long. The female tips in at about a million times the male's weight. In human dimensions, that would be like a woman carrying around a husband the size of a mole.

Mating is further complicated by the fact that these fish live where it is so deep there is virtually no light, making it difficult for them to even find one another. But when, using scent, the male finds the gigantic female, he doesn't risk losing her. He bites into her side and holds on. For the rest of his life. From the moment he bites into her until they die, the male and female are literally inseparable. The area around his mouth melds into her, becoming part of her body. Before biting her, the male had large eyes, but now they begin to degenerate until he becomes blind. Eventually his internal organs are replaced by an extension of her circulatory system. In many ways, he is nothing more than a fetus. With one exception: Although he loses his eyes and digestive organs, his sexual organs grow until he reaches a length of about four inches; in essence his body is nothing more than a holster for his sexual organs. It is thought that the female is then able, at will, to force the male to release sperm. In essence he is merely another appendage now, like a hand or a foot. In fact, females often collect a number of male appendages. Three or four males will attach themselves to a single female. It's like those perfume ads a few years ago: "I don't know where she ends and I begin." (See also SEX/DANGEROUS: BRISTLE WORMS for castration behavior.)

ANT LIONS

The adult ant lion looks like a grasshopper with moth wings. The females have a voracious appetite that includes just about any fly or insect that they see, including their own species. Not only do they devour their mate after sex, but they then go about eating any other males who come near them. And because they only need to mate once, these other males don't even get the thrill of sex before dying.

➾**FYI**: The ant lion, like the **praying mantis**, will continue to eat as long as there is a supply of food available. They are virtual eating machines. She regards anything that she can conquer as food, and since she does not have the physical ability to distinguish between her mate and all other potential food, she eats him. It's as if you always felt as though you hadn't eaten in three days and if you didn't eat soon you'd die; combine that with the idea that every time you looked at another person you literally saw a sandwich. That's what life is like for the female praying mantis and ant lion.

BEDBUGS

This tiny creature performs one of the most violent forms of reproduction in the animal world. Even though the male has a penis and the female a vagina, the male does not insert his penis into the vagina. He climbs onto the female's back in a position where his penis couldn't possibly reach her vagina; even if it could, it's too big to actually fit. So he improvises: He jabs the sharp point of his penis directly through her back, penetrating the tough outer body covering and ejaculating into the tissue called the Organ of Berlese. When the tissue is saturated with sperm, the sperm break off and travel to storage pouches where they wait until the female has her next meal of blood. The blood stimulates the sperm and they once again begin traveling, this time to the ovaries in order to fertilize the eggs. The storage of sperm is crucial, because too much sex will kill her; six times (and six stabbings) can be lethal. In fact, sometimes during sex the male bedbug stabs the female in a fatal spot, killing her. (See also HOMOSEXUALITY: BEDBUGS.)

BEES

The queen bee is the most focused, no-nonsense animal alive. From the moment of birth she knows what she must do and she does it without doubts or hesitation. Her job is to lay eggs and she's the only one in a hive permitted to do so. As soon as she emerges from her cell, her first order of business is to assassinate any other bees who are developing into queens. If any have developed to adulthood at the same time as she, they fight to the death. Next order of business for the survivor: Get pregnant! She circles the hive then speeds off with all the drones (male bees) in hot pursuit. Details vary from species to species, but in some the female mates only once in her lifetime (but remains fertilized for the rest of her life, about five years). Her only lover is the first drone to catch her in midair. (In some cases, she simply flies straight up into the air, with drones exhausting themselves and plunging to their deaths until she finally mates with the last drone left—the strongest.) Once the drone has thrust his penis into her and ejaculated, we can only hope it was good for him because that will be his farewell gesture. As he dismounts, his penis remains lodged inside the queen and his body is ripped apart, scattering his internal organs everywhere. His penis remains as a plug to prevent any of the sperm from escaping from her. Some queens will be able to fertilize eggs for the rest of their lives with this sperm. Which raises an interesting question for the rest of the hive: What do we need males for now? Concluding that they don't, the females kill the remaining drones. The males' bodies are torn and sawed apart. Even the drones who may escape don't know where else to go at night, so they return to the hive only to be executed by their sisters.

➾**FYI 1:** How do you get to be queen bee? Actually, the queen is not genetically any different from her sisters. The female larvae all begin life the same, with the exception that the queen's cell in the hive is larger than those of the others. "Nurse bees" feed all the larvae a special secretion from their facial glands, but the queen bee is fed longer than the others and therefore grows bigger.

➾**FYI 2:** Just so you don't think it's only the female bees that are brutal, the male **digger bee** can be equally ruthless. These bees develop underground, but the male is able to dig his way to the surface first. Once above ground, he begins looking for females still underground. When he finds one, he digs her up and proceeds to try to mate with her. However, this is not a society of finders keepers, for as soon as he unearths his treasure, the other males swarm on her in a churning riot of fighting. Dozens of males battle until many lie dead and others are chased away, leaving only one dominant male who instantly grabs the stunned female and flies off with her to mate. (For similar behavior, see also INCEST: MOTH MITES.)

BRISTLE WORMS

This is another one of those horror stories that give men the willies (see also in this section, PRAYING MANTIS and ANGLERFISH).

In general, the marine bristle worm seems pretty harmless, even festive. During mating season, the males perform a lively and erotic dance for an audience of females, whipping them into a sexual delirium that causes them to drop their eggs. Immediately the males soak the eggs with their own sperm.

However, the *Platynersis* genus female is not content to merely drop her eggs. She becomes so sexually frenzied that she bites off the male's sex organ. Though inconvenient for the male, this method of sex fulfills its function: The sperm she swallows along with his organ travels through her body and fertilizes her eggs, resulting in pregnancy.

CAMELS

Camels have a reputation for nastiness toward humans, biting and spitting at them regardless of the amount of love and affection shown toward them. They don't seem to like each other any better. As parents they show little interest in their offspring. As sexual partners, they show even less interest. As with other species, the male and female demonstrate desire by biting, kicking, and smacking each other with their heads. When the male has finally worn the female down, he mounts her from behind—not for intercourse—but to drive her to the ground. He then sits behind her, extends his penis into her vagina, and thrusts for about fifteen minutes. There is no moaning or eye rolling when he ejaculates, no outward sign at all. When he's done he merely rises and goes off to eat, ignoring his recent conquest altogether.

➾**FYI:** According to Robert A. Wallace (*How They Do It*), camels may have the distinction of being the first animals to use an IUD. Camel herders used to place apricot pits in the uterus of the female camel to prevent them from becoming pregnant on long caravans.

CATS

Sex among cats is a brief but rough affair. During breeding season, the female gives off a scent that drives the males wild with passion. A male will suddenly become obsessed with a female he might have ignored last week. Once she raises her rear, he is on her. He clamps her neck between his teeth and jams his penis into her with several quick thrusts. That's pretty much it. Except that his penis is covered with backwards horny barbs and withdrawal can be a dangerous time. When she tries to roll onto her back to dislodge him, he must leap backwards, far enough out of her reach to avoid the sudden attack she is liable to make at him. Despite the after-sex animosity, the female will be ready for another round in about an hour. This will continue for about four days, during which time the male will become thin and exhausted from not eating.

CICHLIDS

This small exotic fish is a case in which the courtship can be more dangerous than the sex, especially if the female doesn't turn out to be the dream date the male had hoped for. The male is an aggressive animal to start with: He digs out a little nest at the bottom of a stream and defends his territory against any intruders except pregnant females. If another male wanders into his area, the cichlid will attack until the usurper either swims away or shows submission by allowing his bright colors to fade to drabness and tucking in his fins. Only then will the male defender stop his attack and let the other male go on his way.

When a willing female enters his territory they engage in a courtship that tests her worthiness, not unlike the football quiz Steve Guttenberg gives his fiancée in the movie *Diner* to see if they are compatible for marriage. The test consists displaying their colors, spreading their fins, and swimming parallel to each other like syncopated swimmers. If she passes the test, he lets her lay her eggs in his nest and he fertilizes them with his sperm. If, however, the female is unable to swim fast enough (not uncommon since the male is larger) she then swims away from him as fast as she can. The male may pursue her, even to the point of killing her.

FLEAS

The male flea's penis is such a spiny, hooked appendage that it requires considerable force to insert it, often resulting in serious injury to the female. To counter this assault, some males also have an organ shaped like a feather duster, which they use to gently stroke and soothe the female during sex.

FLIES

The **robber fly** is a large and dangerous predator, with the skill and speed to gobble up pretty much any insect it comes across. This predatory disposition makes it dangerous to mate; the male is well aware that if he gets close enough, the female may choose to eat him. Therefore, he merely waits until she's preoccupied with eating a meal, hops on her, and has his way. As for her, she continues feeding, pretty much ignoring him. (See also PROSTITUTION: FLIES for more creative ways that the female is distracted.)

Dance flies are even more deadly. Most of the 2,800 species are ravenous carnivores who'd eat their own underwear if they wore any. Naturally, such an indiscriminate appetite makes mating a chancy enterprise. The male, which is smaller than the female, increases his odds of surviving sex without being eaten by his mate by presenting her with the gift of an insect. The gift does not appease her aggression; it merely distracts her while he quickly mates with her and then flees. His gift-giving technique is unusual. Rather than just dump it in front of her, he holds his gift in front of him like a shield and flies—along with a gang of other males equally prepared—into a swarm of female dance flies. The females pounce on the gifts and all fall to the ground. While the females munch away, the males have their way.

Many of the dance fly species in North America first gift wrap their insects with silk from their own glands. In some cases the packaging is so elaborate, the female spends more time unwrapping, allowing the male more time to have sex and escape. The males of the *Hilara sartor* species found in the Alps, have a much more Continental approach to gift giving. Nothing so crass as an insect; rather they give a white veil, which the male has spun himself and which he displays to the female between his four middle and posterior legs. The female is quite content with this gift, even though it is clearly not food. While she examines it, he mates with her.

The female is not the only dangerous one in fly relationships. The more frequently the female **fruit fly** has sex, the quicker she dies. This is the result of poisonous additives males secrete in order to kill the sperm of other male flies who have already had sex with her. (See also BEDBUGS above.)

FROGS

Different frog species have different sex habits. Some copulate the same as humans, complete with orgasm; others wait for the female to eject her eggs before fertilizing them (see also PARENTING: FROGS). However, male frogs can become very intense in their drive to breed. They sing and sing (croaking to us) for hours and if no female responds, their sexual frustration builds. Sometimes they get so horny they'll leap on anything from wood to small animals to dead frogs to other males.

Sometimes they just rape the first female they find. If the male does get lucky and finds a receptive female, other males may attack him while he's mounted on top of her. It is not unusual to see a female frog clasped tightly between two males as each tries to fertilize her emerging eggs. Unfortunately, as each male battles the other for the right to fertilize the female's eggs, they sometimes fight so violently that the female is injured or killed.

➾**FYI 1**: The **African clawed toad** also climbs upon the back of the female in order to squeeze out her eggs, but because he is so much smaller than she, this method often doesn't work. After a frustrating ride, he will climb off her back, grab her firmly, and spin her around until she's so dizzy she becomes unconscious. While she is out cold, the male squeezes her until the eggs are released.

➾**FYI 2**: Sex between some frogs can take quite a while because the male doesn't have a penis. This simple fact changes the balance of power as to who determines the outcome of this copulatory effort. The female releases the eggs in her own good time, sometimes postponing it so long she is forced to hop around with the male still riding her back in anticipation. However, he "encourages" her to release her eggs faster by continually tightening his grip around her middle, squeezing the eggs toward expulsion. When he feels her abdominal contractions signaling that the eggs are about to be discharged, he readies himself to ejaculate on them as they emerge.

➾**FYI 3**: The ancient myth of Pygmalion—whose statue of a woman was so beautiful that he fell in love with it—has been told and retold many times, including in *My Fair Lady*. The myth is a warning to humans not to admire what we create too much because it leads to false pride and removal from realizing that we ourselves are but creations of a greater being or natural force. Such pride traditionally results in the punishment of death—or symbolic death by isolating the offending person from society. This is the perfect myth for frogs because in their frenzy to "love," the males sometimes leap on a lump of mud because it looks and feels like a female frog. They squeeze and squeeze, which causes the lump to resemble the female frog even more. The male will hang on to this statue for days before moving on. Loving his own creation can prevent him from fertilizing eggs and thereby passing on his own genes—which is the death of his line.

MIDGES

Like the **ant lion** and **praying mantis**, the female midge can be a voracious predator, completely indifferent to the lives of the males of her species. Even so, the males go through a lot of effort to attract these females. They gather over ditches, beneath trees and bushes, even around cows and people in dense groups that resemble a hovering cloud. With some species, the females will quickly join the treading males for mating. But in other species—those that hunt other gnats and flies—the females will descend on the unsuspecting males like a horde of savage pillagers. Once a female enters the cloud of males, she will single out one, land on his back, and throw a scissors lock on him with her legs. Then she injects his body, beneath his armor plating, with her saliva, a powerful corrosive substance that soon turns his innards into a liquid that she slurps up like a milkshake. When she's finished, she tosses aside the empty husk like a drained Coke can.

But all is not lost—genetically speaking—for the male. Even as the sated female kicks away his husk, the tip of his remains, which contains his sex organs, continues to stick to her abdomen and fertilize her. He parents posthumously even while his liquefied body nourishes the mother and developing offspring. For those who find the female midge a bit too wicked, she does receive some serious payback at the jaws of her own offspring. (See also PARENTING: MIDGES.)

PORCUPINES

Those long pointy spines protruding all over the porcupine's body are just as dangerous to other porcupines as they are to humans and other animals. That makes sex a very delicate maneuver worthy of the *Mission Impossible* crew. (See MASTURBATION: PORCUPINES and SEX/KINKY: PORCUPINES for a description of the prelude to intercourse.) Once the amorous couple is ready, the female may flatten her prickly spines and twitch her tail to the side, giving the male free access. Or she may stick her tail straight up and walk backwards to meet him. He approaches on his hind legs, thrusting his pelvis at her genital area, which is covered by nonthreatening soft hair. However, if her tail is not positioned correctly, her sharp spines could stab him. The young porcupine is especially at risk, because his penis is not as long as the older porcupine. This is where age and experience have the decided advantage over youth and enthusiasm. Not only is the more mature porcupine endowed with a longer penis, but he knows how to brush aside the tail just right to reduce the danger. The male thrusting is unusual in that he may not touch her at all, holding his paws in the air while pushing his tail against the ground to give him the rocking momentum necessary to ejaculate. The female may thrust her rear against him, which is why sex is so brief. The instant he ejaculates, the male withdraws and begins licking his own penis or takes it in his mouth.

➾**FYI:** The couple continue to have sex, with twenty-minute rest periods between, until the male is too exhausted to perform. He then wanders away alone.

PRAYING MANTISES

This is the one that scares all the guys. A Freudian nightmare. Once you've actually witnessed it, you're not likely to forget it; nor will you be able to wax poetic on the beauty of nature with quite the same abandon as before—unless you find efficiency beautiful. If so, this is the creature for you. The praying mantis is a startlingly fast and powerful insect that is capable of catching and eating frogs and small birds. It's a kick-ass bug and it takes no prisoners, even among its own.

As is often the case with insects, the male is smaller than the female and must approach her with extreme caution. In fact, he usually waits until she is doing something else, such as capturing another insect, before sneaking up on her. He knows that if she sees him within her striking range, she will kill him. But he's descended from a long line of males who took the risk, because those who were faint of heart never got to pass along their cautious genes. When he does finally hop onto her back, they begin having sex (which might be right side up or even upside down while hanging from a branch). He ejaculates sperm, which she then stores until she wants to be pregnant. When he's finished, he spreads his wings and flaps off to safety. If he's lucky.

Even the unlucky still get to have sex—they just don't get to enjoy it. If during his approach the male gets clumsy, the female reaches out, grabs him, and shoves him underneath her. Sex has been known to produce a powerful appetite; but there are no midnight raids of the refrigerator for this couple. In anticipation of sex, she just starts snacking on *him*. She begins with the eyes, then the head. As she chews through the head and into the neck, she demolishes the nerves that inhibit sexual behavior. That means that the male is now the kind of stud machine that he never would have been while still wearing his head. It is now that his body writhes and rotates so that his penis enters her and he pumps away like the Don Juan of praying mantises. She continues to munch away on him, even as he is ejaculating sperm into her. It ends only after she has eaten the part of his body to which the penis is attached. She leaves only the wings behind.

➾**FYI**: In one species of praying mantis, the male's brain is hardwired in such a way that it cannot release its sperm during sex. The only way for the sperm to ejaculate into the female then is for the head to be removed. In this case, the female must chomp off the head if there is any hope of the species surviving.

RHINOCEROSES

If you've ever seen wildlife footage of an enraged three-thousand-pound rhino charging a fleeing Jeep, you can understand how sex between such powerful creatures is so risky that on occasion one of them is killed during the act. In fact, when you think of sex between rhinos, recall the image of the rhino charging a Jeep, the metal door crumbling under the massive impact, the whole car listing to one side. Because that's basically what it's like.

At first, the mating ritual is fairly innocent: The female comes into heat, her vulva swells, she loses her appetite, she makes shrill whistling sounds. The male answers back with deep sighs. Then both begin to act like they're on a blind date gone bad; they toss their heads and pace anxiously. Then the male charges directly at the female. But she is no shrinking violet—she charges right back at him. If this is a game of chicken, neither balks. They crash head-on. And they keep doing it over and over, butting and dueling with their four-foot horns. After about an hour, they become exhausted, which is when they are most vulnerable to injury. It is at this time the female accepts or rejects her combatant as her lover.

Now the danger increases. The male mounts the female from behind, but he is not as skilled at sex as he is at fighting. Although his penis is two feet long, he has a lot of trouble actually placing it into her vagina. A whole lotta shakin' and thrustin' goes on before penetration, and even then he may only have managed to poke a couple inches inside. However, what he lacks in finesse he makes up for in determination; he remains mounted, ejaculating every ten minutes. And each time, he climbs further onto her back, thrusting more and more of his penis into her. By the time he's finished, all three thousand pounds of him is completely off the ground and on her back. When he's finished and withdraws, about an hour later, both are completely exhausted. Yet, it is not uncommon for them to do it all over again an hour later.

➾**FYI**: The risk of death or injury is so great during sex that zoos are cautious about breeding rhinoceroses for fear of losing one. One reason for this dangerous foreplay is that rhinos are not very social creatures. They are solitary animals who steer clear of each other in order to avoid fighting. But when mating season arrives, they are forced into each other's company, igniting their antisocial behavior. The impulse to fight and the impulse to mate struggle within them until one is seriously injured, dead, or the mating impulse wins.

SNAILS

Next time you're chomping away on *escargot* (the **land snail**), remember that its sex organs are located in its head. Because each snail contains both male and female sex organs (see SEX/GENDER BENDER: SNAILS and SEX/GROUP: SNAILS), when they engage in sex, both partners ejaculate sperm into the other, thereby fertilizing each other's eggs. Strange to us, but routine to many species. What is startling is the level of violence involved during sex. At first they touch each other gently and slowly; eventually they rear up and stroke each other with their sensory stalks, not unlike humans kissing. Suddenly each thrusts an inch-and-a-half appendage shaped like an arrowhead into the body of the other—perhaps the heart, lung, or brain—sometimes instantly killing its partner. However, if both are still alive, they seem stimulated by the pain of the stabbing. They separate, then come together again. Each slides its penis into the vagina of the other, ejaculates immediately, then separates again. They lie there without moving for a while as they recover, then inch away in opposite directions. (See also BEDBUGS above.)

➾**FYI 1:** The so-called "war between the sexes" that is so often argued over among humans has produced many fantasies in literature in which one sex is completely self-reliant; that is, it can do everything itself, without the need for the opposite sex. Including propagation. With the *Rarae aves*, this fantasy becomes reality. Unlike some other animals that internally fertilize their own eggs, or clone themselves, this species of snail actually has sexual intercourse with itself. It inserts its penis into its own vagina and impregnates itself.

➾**FYI 2:** There are over thirty thousand known species of snails, with almost as many variations of behavior, so you will see them mentioned throughout this book under different categories.

SNAKES

This is another cautionary tale for males and females. The weirdness starts right when the male and female, who identify each other by smell, begin to tie themselves into a love knot. While this knot may just be the two of them, it often also involves one or more additional snakes. The belief is that the male does this in order to not only hold onto his love object, but also to tie up the male rivals so they can't interfere or have sex with her themselves. It is unknown whether the female could escape or whether she is being raped.

Once the knot is tied, the male attempts to insert his penis into the female. However, the female is often not yet sexually mature enough to mate, so the male stops trying to have sex with her. But he keeps her tied up for hours or even days, periodically trying to have intercourse until she finally is ready. When this happens, he inserts his penis, a spiny, bumpy, hooked organ that once inside is difficult to withdraw. Once his sperm has been released, they may just lie there awhile without moving. But if a predator comes along and spooks them, the female may slither away dragging the male along beside her by his penis. If the two try to take off in different directions, the male's penis may simply break off. Fortunately, he has another one.

➾**FYI:** Females of some snake species are able to store the sperm in their bodies for as long as five years, fertilizing themselves each year with it. (See BEES above.)

SPIDERS

In some species, spider sex can be risky because the females are not only much larger than the males, but they are also cannibalistic. This dangerous behavior is not the typical behavior, however; among the over thirty thousand species of spider, only a small fraction practice cannibalistic mating. The others practice safe sex. For example, the **ground spider** is pretty amiable and will offer her vagina to the male without any threat to him. The male **crab spider** also engages in some tender foreplay without fear of being eaten by his lover. However, there are some species in which mating means the male risks being slaughtered and eaten by his mate.

But for those who do practice cannibalistic mating, practice has made perfect. The female's appetite is aided by the fact that the physical logistics of sex between spiders is tricky: Male and female have their genital orifices on their abdomens, and the male has no penis. Instead of using a penis, the male spider spins a small web that he uses like a sponge to absorb semen from his sex organ on his abdomen. He then uses the semen-soaked web to fill the little vessels on his feelers (or his pedipalp, the leg with the sperm). When—or if—he is able to get close enough, he sticks his semen-laden feelers into the female's genital orifice (her epigynum) and releases the semen. Now he must pull out and run like hell to avoid being killed and eaten. Sometimes he panics and flees too abruptly, breaking off his sex organ and leaving it behind inside her. (For additional tales of castration, see SNAKES above and RAPE: OCTOPUSES.)

In the *Araneus pallidus* species, the male scoots under the female—working as fast as a safe cracker who's just set off an alarm. However, the position for sex requires that his abdomen be positioned right under her mandibles. She then clamps her jaws around his body, not to kill him, but to hold him in place so he doesn't slip off her slick armor-plated body. The male, held securely by his mate's enormous jaws, begins to copulate with her. This may continue for about ten minutes. Afterward, still clutching him in her jaws, she chomps down and eats him.

The **Australian redback** spider offers an interesting variation: Not only is the male eaten by the female during sex, but he actually has to go to some trouble to *allow* her to do it. In fact, he appears to willingly contribute to his own death—a sexual suicide. It works like this. The male is much smaller than the female, about the size of a penny lying on a dollar bill. Once he has inserted his organ into the female, he somersaults around—

without his organ slipping out—and dangles his abdomen in front of the female like bait. She takes the bait and begins chewing away on him. The reason for this kamikaze sex has baffled scientists, but one explanation offered is that by offering food to his mate, the male can continue his sexual act longer because she's distracted by her meal. This allows him to deposit as much sperm as possible, thereby bettering his chances of passing along his genes. Once a female has been filled with sperm, she tends to rebuff any other suitors. In any event, even if they weren't consumed by the females, the males wouldn't live long after mating anyway. With this method, at least they've improved their chances that the one time they do mate, it will produce the desired effect.

The male **autumn spider** (*Metellina segmentata*) is also at risk during sex, so he has developed a profound patience. He will stand beside a female's web for hours or even days, waiting for some other insect to get caught. As soon as one is, the female spider comes over to eat it. When she takes her first bite, the suitor springs into action, mating with her while she is occupied with eating. The male *Pisauridae* doesn't wait around; he goes out and catches his own insect, then wraps it with his silk until it's a tight little package that requires a lot of patience to open. Then he presents it to the female as a gift. As she's busily unwrapping and then eating it, he scoots in and has sex. If he finishes before she's gotten to the actual insect inside, he grabs the gift and takes off with it.

For some spiders, there's no winning, even if he manages to have sex and escape. The *Araneus diadematus*, a species of European spider, is only a little smaller than the female. When he wants to mate he simply leaps on her like a wrestler, wrapping his legs around her so she can't bite and lasso him with a web. When he's done he scampers away in triumph. But his victory is short lived because he dies of natural causes within several days of mating.

The male **tarantula** does not have very good eyesight, so when he's in the mood for sex, he approaches as a possible mate pretty much anything that is the approximate size of a female. When he finally finds one, he taps his four front feet against her leg or side. Just as she rears up and exposes her fangs to attack him, the male grabs those fangs with special hooks on his front legs. Then he drums on her abdomen until she calms down enough for him to slip his pedipalp into her. After sex is complete, the male can take his time leaving; the female will not attack him after sex.

SWORDTAIL CHARACINS

This relative to the **piranha** produces females that are so hostile to males that mating is a very dangerous undertaking. In order to mate with the female, the male must trick or distract her. Fortunately, he's got just the thing already built in: an appendage with a little crablike knob that he can extend at will. When he sees a female, he opens his gill cover, extends the appendage, and then jerks the gill cover to make the little knob look just like a swimming crab. Then he pretends to flee in fear from her to lower her aggression level toward him. Free to concentrate on the "crab" swimming in front of her, she attacks it and swallows. But she is unable to actually swallow it, though she keeps trying. While she is busy fussing with her food, the male swims up to her and mates.

TIGERS

There are many instances in this chapter of females who devour or attack their mates. And there are many instances in this and other chapters of males who rape, beat, even kill their female mates. The key to who does what to whom is usually size: the bigger they are, the more likely they are to exploit the smaller ones. It's pretty much schoolyard rules: the biggest most aggressive bully takes what he wants. But in human society, we have the ability to neutralize this rule of the jungle. The bully soon realizes that: (1) even a small kid can find a weapon (as in the old saying, God made people, but Smith & Wesson made people equal); and (2) laws try to punish those who would exploit others through physical force. Once a child realizes he can't just take what he wants, he must develop social skills.

However, when an animal can take what it wants, there's no incentive to develop social skills. This is why animals that attack their own mates are usually from nonsocial species, loners with no experience at socializing. No team sports to train them. Yet, there are some cases in which the male animal is much more powerful than the female, but the female is able to attack and sometimes kill the male, who surprisingly puts up no defense.

Such is the case with tigers.

When the tigress comes into heat, she abandons her normally isolationist lifestyle and suddenly starts rolling around on her back acting like a baby. Many animals—including humans—adopt this infant role-playing technique during courtship (see COURTSHIP: SQUIRRELS). It sends out a signal that the animal is harmless. The male tiger, who knows that the females are usually as antisocial as he is, now will approach the female. They spar, nip, and flirt until the female allows the male to mount her. At this point, the male will clamp the nape of her neck between his teeth as he would a prey he's just downed. But instead of chomping through the neck, he merely bites her lightly. Then they mate.

Now is when things really get dangerous—for the male.

After they are finished mating, the male's aggressiveness is numbed; he's in the comfort zone of an afterglow. But the female's aggressiveness returns like a gunshot. She instantly turns on her lover and attacks him. Despite the fact that the male is stronger and could defeat her in a fight, he runs for his life. His feelings of attraction for her dampen his otherwise natural aggression to the point where he cannot defend himself. So he flees. In the old-fashioned zoos where the male and female were kept in small cages, the female would sometimes kill the male immediately after mating.

2. Duration & Frequency

Duration

How long do humans engage in a single sexual coupling? Again, we have statistics, but researchers agree that such figures are distorted by how individuals interpret the beginning of sex. Unmarried couples tend to think it begins with hugging and kissing, while fully clothed. Married couples see sex as starting once they are naked. Also, psychological studies indicate that people think things take longer than they actually do, and tend to overcalculate how long their sex takes. Still, for comparisons' sake: 11 percent say sex takes them less than fifteen minutes; 69 percent say it takes them between fifteen minutes and an hour; 20 percent say it takes them longer than an hour.

Frequency

How often do humans have sex with a partner? According to the most definitive study to date, detailed in Sex in America *(Robert T. Michael, et al.), Americans between the ages of eighteen and fifty-nine have sex less frequently than previously thought. One-third have sex with a partner at least twice a week; a third have sex with a partner a few times a month; and a third have sex with a partner a few times a year or not at all. The average for men is seven times a month, for women six times a month.*

BABOONS

The female baboon is unusual compared to most other primates. Like other primates she will have sex with many partners during estrus; but, unlike the others, she will also have sex when *not* in heat. Sometimes she will even seek sex when she is pregnant, which is rare in nonhuman primates. No matter how often she wants it, the actual sex act usually takes only about seven or eight seconds. What lack there is in duration, she more than makes up for in frequency. One female baboon was observed to have mated ninety-three times with three different males during the five days of her estrous phase.

➾**FYI: Chimpanzees** copulate with the maximum of efficiency. The female crouches and presents her rear, and the male, with no foreplay, enters her. It's all over within ten to fifteen seconds. This allows her to move right on to the next male, who is usually waiting patiently for his turn. (See also SEX/GROUP: CHIMPANZEES.)

ELEPHANTS

The average female human has sex six times a month; the average female elephant has sex *twice in ten years*. It's not that she doesn't have the opportunity; it's just that she's not in the mood. After her first bout of sex, if all goes well and she's pregnant, she has to carry her offspring for twenty-two months (longer than any other mammal). She doesn't want sex while she's pregnant. Nor is she particularly interested in sex for the three years after giving birth, for then she concentrates on raising her young. Once her maternal duties are pretty much wrapped up—after five years—she's in the mood for sex again.

➾**FYI:** During the Middle Ages, monks produced very popular books called bestiaries, illustrated texts based on ancient writings that purported to explain animal behavior. The monks added travelers' stories and legends to these texts so that as time passed they became more outrageous. One feature of the monks' work was to offer explanations of why certain animals existed. The elephant, for example, because of its size and strength, was here as a reminder to humans of God's mighty power. Also, elephants demonstrated the ability of virtuous sexual control—in fact, they symbolized the purity of Adam and Eve before they ate from the forbidden fruit. The monks believed elephants did not mate here on Earth, but in Paradise. They reached this conclusion because travelers who observed elephants in the wild never reported seeing them have sex. We now know that part of the reason is the infrequency of the act—once every five years.

LIONS

A lioness in heat redefines all human notions of passion and promiscuity. Most female mammals ovulate on their own monthly, but the lioness needs sex to ovulate. She is ready for sex every fifteen minutes over a three-day period. One lioness in estrus had been observed over a sixty-hour period to have had sex 170 times, an average of once every twenty minutes. When she wore out her first lover and he staggered off, a fresh lover took over the task. One reason for this insatiability is that the lion cubs frequently die very young. (See also PARENTING: LIONS.)

MICE

The **marsupial mouse** is dedicated to sex to the point of fatality. He will copulate with several different females over a period of up to twelve hours. Afterward, the pregnant females shuffle off on their merry way, but the dazed male simply lies down and dies from exhaustion. Mice sometimes mate as many as seventy times during one lovemaking session, but the **Shaw's jird**, a small Israeli rodent, has been observed to have mated 224 times in two hours.

➾**FYI**: Wild **rats** have been clocked mating four hundred times in ten hours.

SAGE GROUSE

This bird can be found in the northern plains of the United States and has the same general frequency of sexual activity as many of the humans who inhabit that part of the country—about one hundred times a year. The big difference: It has all one hundred sexual encounters in *one night*! The rest of the year he does without.

The circumstances leading up to this big night of activity are crucial. In early spring, three months prior to the actual sex, the males begin to arrive to stake out their breeding ground. For the next few months they march about, preening and fighting with each other as they try a hostile takeover of better real estate. The typical individual space consists of about ten feet square, but there are about five hundred sage cocks. The reason for the fighting: only 1 percent of the males—about five of them—will actually get to breed. It's pretty much like a game of King of the Hill, with the kings being the only ones who will pass along their genes. Although that means most males will not get to breed and their DNA will die out with them, the species itself will be improved because only the strongest, toughest males will pass their genes on.

After a couple months of angling, the fiercest competitors dominate the prime breeding squares. These rulers also have a royal staff: a sub-cock and several guards. When the big night finally arrives, about five hundred hens will flock to the breeding grounds, but all five hundred will want to have sex with only the five cocks that are on the prime real estate. Because of the number of females and the amount of work to be done, there's no direct courtship. The males go right at it, copulating with each hen in turn. However, even a male of proven strength and endurance can weaken under such a heavy demand. When this happens, the hens will often wait for the hour or so it takes him to regain his strength. Still, he may give out before servicing them all. If so, his next in command steps in to continue the copulation. It is even possible that the guards may get to engage in a little sex.

As for the other 495 sage cocks who ended up in the cheap seats as observers, even if they tried to copulate with the hens, the females would not accept them. The females want only the strongest of males to merge with their own DNA, thereby giving their offspring a better chance for survival. (For a similar use of prime real estate as means of establishing who has sex, see COURTSHIP: KOB.)

SPARROWS

The usual rate of intercourse for wild birds during breeding season is once a day. But the common house sparrow couple has been observed having sex up to fourteen times in a day.

3. *Gender Bender*

It has been the plot of many fantasy movies: What if a man and woman switched bodies? Then each would have some idea of what the other was going through. As usual, nature is way ahead of Hollywood.

EARTHWORMS

One aspect of the birth-control debates is that the burden has generally fallen on women to make sure they are protected from unwanted pregnancies. They have often complained that males, in their sexual rush, are too often cavalier in their attitude, and unreliable. "What if men got pregnant?" women have often countered. "Then they'd be more careful." Well, with the earthworm, that's almost what happens. Because earthworms have both male and female genitalia, their copulation results in *both* of them becoming pregnant.

➾**FYI:** Technically, this sexual encounter is purely a homosexual one, since it involves only the male sex organs of each. Each earthworm ejaculates from its erect penis into the special sperm pocket of the other. It isn't until later that this stored sperm is used to fertilize the eggs. (See also HOMOSEXUALITY: EARTHWORMS.)

SEA HORSES

If you're looking for something in nature to show how males are able to handle the responsibilities of pregnancy and newborns as well as females, look no further than to the sea horse. When two sea horses mate, the female grabs the male with her prehensile tail and they sway about some. As they rise to the surface, the male's abdominal sac inflates and the female begins jabbing at his body with her bright orange "penis." (Okay, it's not really a penis but a nipple of flesh; though in essence it serves the same purpose as a penis.) When she finally finds the opening to his pouch, she pushes the nipple in and ejaculates up to six hundred eggs into him. Afterwards, she swims off, footloose (if she had feet) and with no responsibilities, despite leaving the male in the "family way."

The pouch acts like a uterus and fifty days later the male goes through labor, experiencing wrenching contractions. Each contraction spits out another tiny sea horse. If he fails to expel any, the rotting corpse inside his body may kill him.

SNAILS

Human conventional wisdom, as depicted in our literature anyway, generally portrays males as ramblin' types who leave the bosom of their families to seek fame and fortune and adventure. Females are generally portrayed as homebodies (hence the phrase "bosom of family") who seek the stability of home and hearth. It is the woman who tames the wandering male, forces him to mature to manhood and accept his responsibilities to family and society, not just his own selfish wanderlust. The aquatic **slipper snail** embodies this concept in the flesh.

All young slipper snails are males. While males, they wander about. But when they begin to mature—as with many humans—they begin to settle down. For us it may be a house with a white picket fence or a condo overlooking the park; for them they just attach themselves to something stationary. Once they attach themselves to something, they begin to transform into females. So, a young male may be wandering along and decide to attach himself to the female. Before he changes over to female, he has a last fling with his new partner. Now, along comes another young male and attaches himself to the female who is attached to the other female. He has sex with the outermost female and soon also becomes a female. And so it may continue until as many as fourteen snails are attached to each other. Once attached, they remain this way for the rest of their lives.

TAPEWORMS

What could be spookier than being told you have an animal living inside of your intestine that feeds off your body and reproduces, getting bigger and bigger, like the creatures in *Alien*? The tapeworm does just that, and without the benefit of having a mouth or any digestive system of its own. It merely lies in the intestines of animals and absorbs nourishment. How then does it manage to reproduce in there? It simply has sex with itself. The tapeworm is made up of a head, neck, and chain of fifty to one hundred segments, each segment containing nothing but sex organs, including a vagina and penis. Although a penis could fertilize the vagina of its own section, it usually chooses to engage in sex with a more mature segment. To accommodate this, the tapeworm loops around so one segment can reach the other.

4. Group

BACTERIA

For bacteria, sex can be a form of self-defense. In general, bacteria reproduce through fission—dividing themselves into two separate creatures. However, they will also occasionally sidle up next to each other like two shoes in a closet and exchange genetic material through their cell walls. This is asexual sexuality. Each of them is now genetically altered, literally a different being. Humans are often extolling the virtues of their loved ones by explaining how love has made them a different person. Well, in this case that's literal.

The advantage for the bacterium can be survival. For example, if one bacterium that causes typhus develops an immunity to an antibiotic, it can pass that immunity on through this asexual docking procedure. So, in a frenzy of sexual activity, it docks with as many others as it can, inoculating its mates each time.

CALLICEBUS MONKEYS

This small monkey from Columbia differs from most primates (and other mammals) in that it forms lifelong monogamous relationships. The happy couple then claims a piece of territory as its own and aggressively guards it from intrusion by other monkeys. This is a strictly "good fences make good neighbors" society, with animosity rising steadily among the group as they constantly battle each other over property and in defense of their mates. Left unchecked, this building bitterness could result in the Callicebus monkeys destroying their society through war (as humans and some **chimpanzees** do) or cannibalism (as **springtails** and others do). Fortunately, the tension is not left unchecked.

When the females come into heat, it's party time. Or more accurately, orgy time. Mates that were recently fought over with such vigor are now tossed aside as the community engages in active, even frenzied group sex. Petty differences accumulated over the past year are forgotten; the community is reborn and recommitted to each other in a ceremony of lust. But after a few days, when the estrous period of the females ends, so does the sex and therefore the goodwill toward fellow monkey. The couples return to their territory and the squabbling begins anew. Until next year's party.

CHIMPANZEES

It is well-documented that the chimpanzee may be the animal closest to human in terms of physical makeup. Therefore, we are especially interested in their behavior as a means of better understanding our own. Young and old, chimps are in general very promiscuous. Humans, especially adolescents, have a grim ritual known as "pulling a train," which means a group of males have sex with a single female in rapid succession. Though this behavior is generally not socially acceptable among humans, it is not uncommon practice among chimps. A willing female will sometimes mate with every male in their social group. While she is having sex with one, most of the other males in the colony stand around watching—and waiting their turn. When one finishes, another takes over. Even the adolescent males join the line, but at the very end.

In many animals, the female's desire to copulate is governed by her menstrual cycle; the desire increases as her body becomes more receptive to pregnancy. A female chimp menstruates every forty days, with her most sexually receptive time coming toward the end of the cycle, during which the skin around her vagina and anus bloats enormously and turns bright red or pink. Although this may be her most receptive time to become pregnant, her sexual activity is not restricted to only these times. The female chimp often has sex even when she cannot possibly become pregnant. This suggests that sexual activity among the chimps is not just for procreation. Researchers speculate that sexual activity also helps form and maintain social bonds that keep peace within their community.

Perhaps because sex is such a social activity, fellow chimps feel it is okay to interfere with others while they are in the act of copulating. Fearful of interference from the dominant males in a group, the adolescent males avoid having sex in their presence and are often forced to rush the whole process in order to finish while the dominant males are otherwise engaged with grooming or napping.

Adolescent females can become their mothers' own rivals for the affections of males. If both females are in estrus at the same time, the daughter may drag her mother's lover off Mom and try to copulate with the male herself. Amongst humans, in such a situation the younger female generally has the advantage over the older one; but among chimps there appears to be a decided preference among males for the older females. (See also SEX/KINKY: CHIMPANZEES; SEX/DURATION & FREQUENCY: CHIMPANZEES; PARENTING: CHIMPANZEES.)

➾**FYI 1:** Although the duration of the male's sex act is relatively brief—ten to fifteen seconds—he is able to recuperate quickly and usually will copulate several times each day when a female is in estrus. That is why when the female comes to the end of her estrous period, she is usually bruised and exhausted.

➾**FYI 2:** The female chimp is most receptive during estrus; that is when the male is pretty much guaranteed that if he approaches her for sex, she won't refuse. However, when she is not in estrus, the female might be more selective and refuse some suitors. To more accurately determine her mood, the male chimp will insert his finger into the female's vagina, then sniff. If he can't detect the pheromones, he moves on.

GORILLAS

Because of their size and strength, humans often regard gorillas as a symbol of virility. However, the reality is quite different. Though the males are indeed huge—six hundred pounds and seven feet tall—they have very little sexual drive. This could be because the females take nearly four years to raise their young, so they only mate once every few years.

Gorillas live in groups that often consist of a silver-backed male leader, two adult males, half a dozen adult females, a few adolescents, and four or five infants. Among this group, the females may mate with whichever males they choose, usually more than one.

GRUNION

These six-inch silvery fish practice a form of group sex that involves not only several males with each female, but it also involves a certain degree of efficiency so they don't suffocate before it's all over.

First, a little background: Grunion-hunting is a popular southern California activity. Every two weeks between the end of February and the beginning of September, the grunion swim onto the beaches by the hundreds of thousands. Even though they cannot breathe outside the water, this is not a case of mass suicide, like whales beaching themselves—it is mass sex. In the two to three minutes they can survive out of the water, the females burrow down into the sand, tail end first, until one-third of their body is buried beneath the sand and the rest is sticking straight up like Popsicle sticks stuck in the sand. Now the males swarm and writhe around them in groups of two or more, fertilizing the one thousand to three thousand eggs the female lays as she releases them. Afterward, they all wriggle back to the sea where they can finally breathe again.

SNAILS

Snails have both male and female sex organs, so sexual intercourse can involve being female one time and male next time. Or not uncommonly, it can involve being *both at the same time*. Before we get to the orgy part, a little explanation regarding the mechanics of snail sex is necessary. Typical courtship consists of two snails meeting and moving in a circle, usually counterclockwise, like fighters sizing each other up. The snail that catches the other gets to be the male during this sexual encounter (though should they meet again, it could be reversed). The penis unsheathes itself to its enormous size: nearly the length of the snail itself. Once the penis is inserted into the "female," the male instantly ejaculates. They may immediately separate or stay stuck together for hours. Sometimes while two snails are copulating, another snail may wander by and decide to join them. "He" climbs aboard the "male" snail and inserts his penis into the "male" snail's vagina. And another snail may climb aboard him. Thus they end up with a group-sex chain in which the first snail is solely female and the last snail is solely male, but all others between them are acting both as males and females, getting and giving at the same time. (See also SEX/GENDER BENDER: SNAILS.)

➾**FYI:** Though it is unclear exactly how this is done, snails are capable of fertilizing themselves. Technically they are virgin mothers. But giving birth is a more difficult process for them because their vaginas have not been enlarged by a penis. (See SEX ROLES: SNAILS for snails that have sex with themselves.)

5. *Kinky*

BEDTICKS

Remember the eating scene in the movie version of *Tom Jones*? The man and woman stuffing their faces with gluttonous abandon as a prelude to sex. Food and sex have a long history in human literature; both remind the human being that he has desires and needs that are hardwired into his being. As much as he may spiritually want to soar with angels, he must also masticate, procreate, and defecate. That's why we place such a high moral premium on controlling our appetites—it makes us feel as if we are in control of those bodily functions. When we aren't attempting to control them, we may be indulging in them, such as giving chocolates to a date. After all, chocolate produces a chemical reaction similar to the sensation of being in love. Or we may even use food as a substitute for love, as many therapists have diagnosed some cases of overeating.

The bedtick, like the **flea** and **bedbug**, feeds like a vampire off the warm blood of mammals. However, the bedtick indulges its desires more extravagantly than those other insects. It has sex *while* it eats. Once the female has latched on to a blood-donor host, the male jams himself between her and the host animal. He then performs a version of sex that is like some gynecological smuggler. He thrusts his large proboscis into her vagina, expanding the opening. When the vaginal opening is large enough, he uses his proboscis and feelers to insert a packet of sperm. During the entire procedure, the female continues to slurp blood, which actually promotes the fertilization of her eggs.

CARDINAL FISH

Different species of fish have different ways of fertilizing the female's eggs. Some practice copulation, others fertilize the eggs after they've been released by the females. But the female cardinal fish inserts her enlarged genital mound directly into the anus of the male in order to extract his sperm.

CHIMPANZEES

Among humans, age is a significant factor regarding when a person should engage in sex. Sometimes we consider a person too young to give their consent. Also, we have various cultural taboos about age differences between consenting couples; a young man and an old woman (as in the movie *Harold and Maude*) is generally considered a little grim. Same thing with older males and adolescent females. Yet among our closest animal relatives, the chimpanzee, sexual activity is not very exclusive when it comes to age. Elderly female chimps sometimes have to squat very low so that the small penises of the infant males can reach their vaginas. Though infant males routinely copulate with the help of the kindly older female, infant females rarely have sex. However, once the female reaches adolescence, adult males will begin to mount her. (See also SEX/GROUP: CHIMPANZEES and PARENTING: CHIMPANZEES.)

ELEPHANTS

Anal sex is not uncommon among humans, though it is soundly condemned by many religions as a perversion since it cannot result in pregnancy and thus "thwarts nature." However, anal sex does occur often among animals, though sometimes it is the result of a fumbling error by the male searching for the vagina—as is the case among elephants. First, it should be noted that elephant courtship is a tender and loving affair. They kiss by inserting their trunks into each other's mouths; they intertwine their trunks above their heads; they are, in fact, inseparable, eating, sleeping, and traveling together for days, even months, before actually engaging in sexual intercourse. (Although, there is much mutual masturbation as each strokes the other's genitals with their trunks.) When mating finally occurs, it is only through the female's invitation. As with most creatures, this is when things get tricky.

The elephant penis is about four or five feet long, but because of the logistics of sex only a few inches actually ever enter the female. The length is necessary because the location of the female vulva is such that the male has to hook his penis into it. Sex occurs when the male mounts her from behind, his front legs resting on her back or sides. That would be an enormous amount of weight (twelve thousand pounds!) to be shifting about on her back in search of her oddly located vagina. That's why his penis is motile, able to move on its own, without dangerous thrusting and shifting about. The penis searches for the opening, slithering about probing for the vagina. Apparently elephants are not very skilled lovers, and anal intercourse is often the result. However, they repeatedly have sex several times a day for several days until the female no longer wants to; it is assumed that at least a few of those times the vagina was also engaged.

➾**FYI:** An elephant is pregnant for almost two years, and during the next three years spent raising her calf, she has no interest in sex. This means the female elephant has an active sex life only about once every five years. (See also SEX/DURATION & FREQUENCY: ELEPHANTS.)

FRUIT FLIES

Fruit flies are especially prized among the scientific community as subjects of experiments on heredity because of their prodigious reproductive habits. They have sex, lots of it, and produce offspring, lots of them. But before the male actually has sex with the female, he likes to dip his foot in the pond first, test the waters, so to speak. Only this is literal. Because his taste organs are located in his feet, he starts his courtship campaign by tapping his forelegs against the female in order to taste her. This taste test allows him to determine if she is of a compatible species (there are over eighty thousand species of flies, sixteen thousand in the U.S.). When he's satisfied with her breeding, he begins circling her, vibrating his wing that is closest to her. Then he performs a little oral sex—using his foot, since that's where his taste buds are. Afterwards, she spreads her wings and vagina and he mounts her.

KANGAROOS

Close your eyes and imagine your parents having sex. You didn't do it, did you? Because there are some things we just don't want to know about. Of course, in other countries, where they don't have the luxury of homes with many rooms, children are aware of their parents' nocturnal sex and think nothing of it.

Kangaroos are like the latter group, completely uninhibited about sex in front of offspring. (See CHIMPANZEES above for adverse reactions to parental sex.) The female kangaroo can come into heat even if she already has an offspring tucked into her pouch. During the foreplay—which she instigates by bouncing after the male until he gives chase to her—she protects her baby by tightening special muscles that seal the pouch. When she has sex, the male positions himself behind her, grasping her shoulders with his small forepaws. His penis is behind his scrotum and points downward. When he enters her cloaca, he thrusts downward; when he ejaculates, the sperm separates into two groups. This is no accident, because the female's tract also separates into two vaginas. That may seem odd, but not as odd as the fact that she also has a third vagina, one that is closed during sex, but opens only to give birth. (See also PARENTING: KANGAROOS.)

PLATYPUSES

"Candy's dandy, but liquor's quicker." You've probably heard this old saying that refers to one of the best known aids to seduction: drugs. This is especially important if the one partner, usually female, is not so willing. Obviously if one drugs or intoxicates a woman in order to have sex, this is not a relationship based on trust and love—it's someone doing anything to get laid. In essence, it's rape. There's plenty of precedent in the animal world for this type of behavior, with the platypus being a prime example. When a male platypus wants sex, he simply poisons the female.

The platypus is a monotreme, that is an animal with "one hole." The one orifice, called the cloaca, serves all necessary functions, from waste disposal to reproduction. Inside the male cloaca is a bump that is the penis; when aroused, the penis extends out of the hole to permit copulation. The female has no vagina or uterus, but instead has two sperm tubes. Not coincidentally, the male's penis is split into two halves, so it can address each tube. It is held in place by long spines that extend from each tip. When he ejaculates, the sperm exits through many tiny holes at the tips of each half.

That's the mechanics. But the problem is that females are not very amorous and show very little interest in the whole sex thing. The male, fortunately for him, has spurs on his rear legs. These spurs are fed by poison sacs. However, these spurs and poison are rarely used as a defensive weapon. Rather, it is theorized, he uses them to stun the female to allow for more compatible sex. It is also possible that the poison, which is strong enough to kill small animals, may also be something of an aphrodisiac to her. Because the female has no clitoris, this poison may be her only source of stimulation during sex. The poison is only secreted during breeding season.

PORCUPINES

A subculture of human sex is something called "water sports," which involves sexual partners urinating on each other. Many people may wince at this, but it is not uncommon in the animal world, as exemplified by the porcupine. When the male approaches the female during mating season, he stands up on his hind legs and walks toward her, his penis fully unsheathed and erect. When he's within six or seven feet, he soaks her with brief spurts of urine. As impressive as this may be (six or seven feet!), not all females respond well to this treatment. Some leave. Some will stay, but snap and swat at the suitor until he stops urinating. But those in full heat may stay, stand on their own hind legs, and allow the drenching to continue.

➾**FYI:** Male wild **rabbits** and **guinea pigs** also urinate on the female as a gesture of courtship. Sometimes during mating, the female **guinea pig** can be something of a tease, crouching down in her sexual come-hither position, only to scuttle off when the male attempts to mount her. She may continue to do this until he becomes so frustrated, he rears up on his hind legs and urinates on her. Sometimes she finds this arousing and they proceed to have sex; other times, she's just, well, pissed, and proceeds to clean herself.

6. Positions

BATS

Yes, bats do it hanging upside down. Suspended by their toes above a hard cave floor, bats having sex could be tricky and dangerous if it wasn't for the male's unique user-friendly penis. It is curved and can pivot at its base, a little like a gear shift. This means the male doesn't have to do any pelvic thrusting; he just hangs there and lets the penis do all the work. They hang there fixed together for quite a while, mostly because the male's penis is so swollen he can't withdraw; the reason for this is to allow the continual ejaculation of sperm over this entire period of time. It is important for the female to have a large supply of sperm since bats copulate mostly in fall or winter, even though there are no eggs in the female tract yet. The sperm, therefore, is stored to fertilize the eggs when they develop in the spring.

➾**FYI:** Because the sex act takes so long, and is performed during fall and winter when bats are preparing to hibernate, the bats may actually doze off with their sex organs locked together and hibernate in that position throughout the winter.

BEAVERS

Beaver couples, who often remain together for life, engage in the unusual position of copulating while sitting and facing each other.

COCKROACHES

They've been around for 300 million years and there are over 3,500 known species, with nearly as many variations in courtship manners and sexual techniques. The **American roach** is a brusque lover who dispenses with any foreplay and just starts right in copulating the moment he finds a receptive female. The **German roach**, however, is more subtle. He rubs antennae with the female until he's excited enough to copulate. When he is, he turns away from her and lifts his wing, enticing her to climb onto his back. When she does, she will eat from a special glandular secretion. Meanwhile, the male has a bouquet of erections, since he has several hook-shaped penises. Now he begins to back out from under her until one of his penises hooks into the female's tail. Then he scoots the rest of the way from under her and turns around so they are now touching tail end to tail end. The rest of his hooked penises grab onto her tail and they proceed to have sex in this position for up to two hours.

DOLPHINS

Among most dolphins, sex is preceded by gentle foreplay and consummated with a few thrusts and not much fuss. However, the **Ganges dolphin**, a freshwater species, prefers sex that is a bit more of an acrobatic challenge. Right before intercourse, the male and female dive down into the water, turn, and rocket for the surface. Both burst half out of the water, facing each other as their bellies slap together, and grasping each other with their flippers. It is at this moment the male penetrates the female with his penis and they fall onto their sides, thrash about on the surface, then swim away. Don't look for this trick at Sea World.

GORILLAS

People used to believe that humans were the only animal that practiced face-to-face copulation; it was part of the romantic proof that we did not merely have sex, we "made love." Whether or not the conclusion is true, the proof certainly isn't valid, because gorillas also on occasion choose to have sex while facing together (see also BEAVERS above). The female will initiate this position. Just as the male is about to enter her from behind, she may roll onto her back and spread her legs. On even rarer occasions, the male may hoist her completely off the ground and have sex with her standing, her legs scissors around his hips. The usual position is the male behind the female, with her on the ground, arms and knees pressed tight against her chest. However, wild gorillas engage in most of the same positions as adventurous humans: side to side, male on top, female on top, sitting or lying down. A female gorilla in the wild is just as able to initiate sex as the male: If she comes upon a male gorilla lying on his back, she may stroke his penis until he gets an erection, then lower herself onto it. The male gorilla must expend quite a bit of effort to achieve orgasm, sometimes needing a hundred thrusts before he lets out with the scream, roar, or sigh that announces his climax.

The extent of the passion and sexual creativity of gorillas puts Hollywood to shame. In *How They Do It*, author Robert A. Wallace describes a pair of gorillas mating on a sloping hill. The female was on her knees and elbows while the male gripped her by the hips. As he thrust into her, they moved forward down the hill like a couple kids playing wheelbarrow. As they moved forward, she cleared the brush out of the way with her arms. They shuffled along for about ten minutes, their panting and moaning increasing as they went along, until they came to rest against a tree. The female climaxed with short, piercing screams; the male roared. Afterwards, he sat down and she walked back up the hill.

➾**FYI 1:** As with most animals, behavior in the wild is markedly different from behavior in captivity. In zoos, males and females tend to actively masturbate. A female gorilla may also attempt to seduce a male zoo attendant by spreading her legs for him, or fondling her genitals.

➾**FYI 2:** The male gorilla can weigh more than six hundred pounds. But unlike many other animals whose body size usually translates into proportional penis size, the gorilla's penis is only two and a half to three inches long.

ORANGUTANS

This is the same primate we often see in movies (such as Clint Eastwood's *Every Which Way but Loose*). They are about four to five feet tall with long powerful arms able to rip a door off its hinges. Unlike the gorilla, which stays pretty much on the ground, and the chimpanzee, which combines life on the ground and in the trees, the orangutan lives almost exclusively in the trees. And it is in these trees that it mates, sometimes while swinging from a tree branch.

RED ROCK CRABS

The usual rule is that if the male is the aggressor during courtship, the preferred sexual position will be the male on top or behind. This is not the case among the red rock crabs, and for a crucial reason—survival. These crabs are cannibals and one crab will eat any other crab that is smaller than itself. For this reason the crabs form gangs among themselves based on their sizes; those of similar size band together to keep the larger crabs from attacking them. Just like a street gang or an alliance of countries.

Because of this appetite for others of their own species, sex can be dangerous. So, rule number one for a male crab is court only crabs that are the same size as you. The smaller ones he'd eat and the larger ones would eat him. Even after finding the crab that's just the right size, courtship is a hit-and-miss ritual. The male does a series of push-ups, similar to the way he'd behave if he were about to engage in combat with another male. The female may run away. But if she doesn't run, they begin a little dance together, something like an erratic tango, moving back and forth, always about eight inches apart. Eventually, the male raises his two pairs of front legs and the female begins to caress them. If he likes her touch and she likes to touch, they may proceed, though often one of them breaks it off and runs away.

Sometimes it works out. That's when the male slips underneath the female, on his back, like a car mechanic. Despite their amorous intentions, they are both still cannibals. The male at this point is more committed to sex than food, but he doesn't trust himself—or her. Scientists theorize that the reason he assumes this belly-to-belly position is to curb his own cannibalistic instincts, which might surge if he were on top of her in perfect dining position. He also must be concerned about her instincts, because during sex he wraps his legs around her in such a way as to pin her legs and jaw so they can't be used.

SOUTH AMERICAN PHALAROPE

Usually among birds the male is the one with the snazzy feathers. But with the South American phalarope, it's the female who is brightly attired, while the male is drab looking; she is also much larger than he is. This makes for some tricky maneuvering during sex. When the male is ready to mount her, she lowers her body to make it more accessible. However, she is still too large for him to just walk up and mate. He has to lift himself into the air and land on her back. The irony is that it isn't technically even necessary for him to land on her; they could have sex with her mounting him since they don't actually have intercourse. They merely push their cloacae (the multipurpose genital opening that emits urine, feces, sperm, and eggs) together. Apparently, they just prefer sex this way.

WASPS

During mating season, wasps are very sexually active, engaging in intercourse for as long as an hour, sometimes while on the ground, sometimes while flying. Different species also have different ways they like to do it. Generally, the male mounts the female from above, either on the ground or in flight, locking onto her with his legs and jaw. If he jumps on her in flight, he may drive her to the ground before sex. By way of foreplay, he may pet her antennae with his mouth. Then he gives himself an erection by forcing fluids from his abdomen into his penis, which he then inserts into the female's vagina.

This is where different species really prove their differences. The **digger wasps**, for example, may choose to fly off together, alighting from flower to flower, all the time joined together and continually copulating. After insertion of his penis, the male **cicada killer wasp** turns around to face the opposite direction. But because the female is larger and stronger, if the couple is startled during sex, she may fly off, towing him backwards by his penis. However, among other wasp species, the female is smaller, wingless, and nearly blind. In this case, the male tows the female, upside down and backwards, while their genitals are locked together. Not just a love-'em-and-leave-'em guy, he may also offer her a fine dinner by regurgitating his food for her.

WHALES

The blue whale is the largest animal to ever live, so naturally we're a little curious how they manage considering their bulk. How do a couple creatures, each over a hundred tons, get intimate without killing each other? After all, the logistics of the act are the same as for humans: the male inserts his penis into the female's vagina. In general, this tricky maneuver is accomplished by floating on the ocean's surface in a side-by-side position. As anybody who's ever owned a waterbed can attest, sex on water, regardless of the size of the participants, can require some skill. With whales, a sexual nudge could send the other adrift. That's why on occasion a third whale will join in to help. This third whale's job is to lie on the other side of the female—like a headboard—to keep her steady during sex.

Not all whales are slaves to this position, though. Some prefer to mate vertically as they erupt belly-to-belly out of the ocean, standing on their tales. Very athletic, but it's all over within ten seconds. (See also DOLPHINS above.)

➾**FYI:** Some whales have harems of about thirty females, which they sexually service once a year during mating season. Afterward, the male tucks his penis away in a special body pouch until it's needed the following year. The size of the penis—about that of a seven-foot telephone pole—can't just hang down while he's swimming about. Its size would slow him down too much.

7. *Weird*

Sometimes there's just no other word for what happens in nature than "weird." It's not a judgment. It's just that some things are so mouth-gapingly bizarre, that it's hard to remember that human sex is any way related to what we are observing.

DOGS

Current psychological theory tends to dismiss Freud's theory of penis envy, but certainly the dog is enough to inspire penis envy—among men. Throughout history, men have tried many bizarre treatments, exercises, and ointments to make their penises harder and bigger. But the male dog doesn't have to face this problem: He is never impotent. This is because, along with the tissue that makes up his (and most mammals' including man's) penis, the male dog (and **wolf**) also has a bone in there that helps maintain the erection. But as much as this can be a boon, it can also be a bit of an irritant. While copulating, the tissue in his penis swells considerably, so much so that after ejaculating the male dog is unable to extract his penis from the female for anywhere from a few minutes to an hour. The male and female may struggle to separate, but they cannot. The reason is quite practical—ejaculation takes a long time and this prevents him from interrupting the flow of semen just because the orgasm is over. An added bonus: The penis acts like a plug preventing the escape of any of the semen.

EELWORMS

When men and women wish to be especially nasty to the other, they poke fun at the size of the other's sex organ. The female will suggest that the male's penis is too small; the male will declare the female's vagina is too big. (As an example, upon her divorce from Tom Arnold, Roseanne told an interviewer, "We were trying to get pregnant, but I forgot one of us had to have a penis." To which Tom Arnold replied in a subsequent interview, "When we were married, she used to talk about how big it was. Anyway, things change. And, like I say, even a 747 looks small when it lands in the Grand Canyon.") Which leads us to the female eelworm, whose vagina makes our remarks about genitalia size as puny as the minds that conjure them.

We must first distinguish between the different kinds of eelworms (sometimes also called **threadworms**). They are part of the class, *Nematoda* (in which there are about thirteen thousand species). In general, they are tiny beings, between 0.004 to 0.06 inches long, barely visible. The **turnip eelworm** burrows in just under the skin of the turnip; when the female is ready for sex, she first must break through the rind. However, they don't actually come out of their little hole—they merely thrust their vaginas out. The males then come crawling out of their little holes under the turnip skin and begin roving about the surface of the turnip, inserting their forked-hook appendage into whichever vaginas they fancy, and filling them up with sperm. They must seem like gardeners watering a field of thirsty flowers. Once "watered," the female remains in her little burrow while the offspring develop; then she dies entombed in the turnip.

Even more startling is the **bumblebee eelworm**. The male and female copulate in the damp earth, after which the male dies. The female, however, burrows through the earth in search of hibernating bumblebees. Upon discovering one, she penetrates the bee's body and works her way into its tissue. Now that she's all snuggled in, her body begins an unusual metamorphosis: *Her vagina starts to grow larger*. As in those old sci-fi movies of the fifties, it continues growing, absorbing the ovaries and uterus; it grows until it is about twenty thousand times larger than the rest of the worm's body. When it has grown to full maturity, it literally severs itself from the rest of the body and lives as a separate being. It takes its nourishment from the host body of the bumble bee, feeding off it until the worm's offspring are born. In the autumn, the baby worms emerge from the dead bee; the mother remains behind to die inside her drained host.

EUROPEAN GRAYLING FISH

There's a scene in the Trevanian novel *Shibumi* in which the hero is so spiritually advanced that he is able to bring a woman across the room to an orgasm without ever touching her. Of course some would marvel at such an accomplishment, while others would say, "What's the point?" Sex without touching seems to remove the best part. But that's exactly how many species mate, including this fish.

First, the female must dig a spawning hole. Meanwhile, the males go through the usual fighting amongst themselves to see who gets to procreate. The victorious males swim up beside the females, each parallel to the other like a couple ships swapping supplies in mid-voyage. The male begins to vibrate his body; then the female, about an inch away, does the same. After a while, they simultaneously release egg and sperm, both swirling together. The sperm fertilizes the eggs, after which the eggs drift down to the safety of the spawning holes the females dug earlier. That's safe sex.

➾**FYI:** This sexual activity is even less romantic than it already seems, because the female isn't really mating with a male—she's mating with the concept of a male, with malelike behavior. Researchers studying grayling behavior placed an oar in a stream where the females had just dug their spawning holes. Then they began to gently shake the handle of the oar causing the blade beneath the surface to vibrate. The female who had just dug her spawning hole beneath where the oar now was, swam up to the vibrating oar, hovered, then also began to tremble, releasing three to six thousand eggs. In essence, to her anything that vibrates is a male.

PLANT LICE

This is an animal in a hurry. They have to be; they face virtual extinction every winter when the cold kills every last one of them. So in the spring and summer, in the name of efficiency, they don't have time to woo and court and mate. Instead, the females—each and every one a virgin—just goes ahead and gives birth to living offspring. Each of these young is an exact duplicate of the mother, all daughters and all look like Mom.

Then autumn comes and things cool down. Now the females give birth again, this time to males. The males quickly mate with the females from the last delivery. These females, not virgins like their mothers, lay eggs. When winter comes, all the plant lice die. All that survives are the eggs, which hatch in the spring and begin the process again.

13
SEX CHANGES

Field notes. . .

All humans begin life as females. The eventual release of hormones determines which embryos remain females and which will develop into males. For thousands of years people have tried to influence the outcome by various remedies, undoubtedly getting their ideas from the animal world around them in which some animals are able to produce a certain sex at will, or produce only a certain sex, or have the sex of their offspring determined by outside influences, such as weather. Undoubtedly scientists will one day discover how to insure that the birth of a child is the sex the parents want, just as they can now surgically change an adult male into a female, or female into a male.

Of course, many animals already have the ability to change their sex, some at will, others depending upon the social or environmental conditions around them. (The sexual transformation of an animal from female to male is called protogyny; from male to female is called proterandry.) This ability plays havoc with our notions of the natural differences between the sexes, especially since much of the history of human morality has been based on continually evolving definitions of what it is to be male or female and the superiority of one over the other. When an individual has the power to be either or both, those notions of sexual superiority require some fine-tuning. According to Vitus B. Droescher (*They Love and Kill*), "In the beginning, our planet must have been populated solely by females. . . . Thus, much as the idea may offend male pride, Eve was not really created from one of Adam's ribs. Instead, it was the other way around. . . . In fact, we might say that the existence of the male, rather than the eating of the forbidden fruit, was what brought about the loss of Paradise."

This chapter focuses on those animals that can change their sexes. The SEX ROLES chapter focuses on the variety of roles males and females play in different animal societies.

BELTED SANDFISH

Among the species of fish *Serranus subligarius*, the role of male and female shifts back and forth. The male is easy to distinguish: bright orange with dark blue spots and white stripes; the female is a darker, drabber color. When the two sexes mate, the male swims up beside her and begins vibrating. Immediately the female shoots out her eggs. The male covers them with his own milky secretion, fertilizing them. Okay, everybody's done their job; the demands of nature should have been met. But no, there's more. The male instantly begins to lose his bright color, turning darker and drabber. However, the female just as quickly becomes brighter and more brilliant—until now she's the same orange and blue that the male had been seconds ago. Now the male expels his eggs and the female fertilizes them. In fact, the male has become a female and the female has become a male.

There's another advantage to this ability to change sexes

besides being able to fertilize each other. If the female can't find a suitable male, she can lay her eggs, change sex, then fertilize them her/himself.

➾**FYI:** Hermaphroditism among fish families is not uncommon. Those possessing it include **sea bass**, **wrasses**, and **sea breams**.

FROGS

The fear of castration is a lifelong fear for many men; without fully functioning sexual equipment, a man may see himself as expendable to society. The inability to impregnate or to become pregnant are traditional grounds for divorce among many religions and societies. However, among frogs if a male is castrated, he simply transforms himself into a female. On the other hand, a female who realizes the food supply is scarce and any offspring she might have would starve, just wills herself into becoming a male.

➾**FYI:** It is not unusual among frogs and toads to be born a hermaphrodite (possessing both male and female sexual organs). After a while, the hermaphrodite frog becomes one sex or the other, usually male. However, some species of toads retain the ability to switch sexes. If the male is castrated or his testes are seriously damaged, his latent female sex organs begin to develop and he will transform into a fully functional female.

GOBIES

In November of 1995, newspapers and TV newscasts were reporting a phenomenal new discovery in the animal world: a fish that was able to change its sex. That wasn't the remarkable part, since for over two thousand years scientists have been aware of this ability among certain fish. However, it had always been thought that once this gender change was made, it was irreversible—same as if a transsexual human had an operation to go from man to woman. But for a variant of the goby found off the coast of Okinawa and known as *Trimma okinawae*, this is not the case at all. These fish are able to transform themselves from male to female and then back again to male.

About two inches in length, the goby lives in groups of one dominant male and several females. The male is not only the impregnator, but also the primary caregiver for the group's offspring. But if a larger male invades the group, the smaller original male simply transforms himself into a female. As a female, she abandons her former male tendencies of aggression toward other males, and she no longer cares for the young. Harmony and order is maintained. And if the male of a group dies, the largest female transforms into a male.

The transformation takes about four days, during which time the male's testicles wither away and its vestigial ovary starts to produce eggs. The penislike appendage inverts, allowing it to better grasp the eggs and sperm.

➾**FYI:** Researchers have been able to induce this sex change within the laboratory, allowing them to study the changes in the brain that occur during transformation. The affected region of the brain is the ventral forebrain, specifically a group of cells that produce the hormone arginine vasotocin. This hormone directly controls reproductive and parenting behavior of each sex in a wide variety of vertebrates. Scientists believe this is the same part of the brain that dictates transsexuality among humans. A 1995 research report from the Netherlands revealed that the part of the ventral forebrain in humans called the BSTc is larger in men than women. But among men who feel they are really women trapped within a male body, this region is even smaller than in women.

GUPPIES

Female guppies outnumber males two to one and if for some reason too many young guppies are born, they'll eat the young ones in a pattern that maintains that two-to-one ratio. Females—at about two inches in length—are also twice as large as males, although they have drab coloring compared to the more flamboyant males. This difference in appearance is because guppies, unlike most fish, have sex through copulation. The male doesn't have an actual penis—he has a groove etched into his underbelly leading to his anal fins; this is the route his sperm will travel to the female's genital opening.

The problem with this form of intimate sexuality is that the male guppy has to get very close to the female for this sexual exchange. Unfortunately, males can't really distinguish what the females look like—not just the difference between girls and boys, but the difference between its own species and that of others. This sometimes leads it to try to court predators of another species, which results in his date being cut short when he's ingested. These gender-identificationly challenged males are sometimes eaten en mass, leaving few or no males in the guppy community.

When that happens, the females have a few tricks up their fins to help ensure the survival of the guppy way of life. First, they are able to store sperm for up to nine months; this allows a female to spawn eight times for a possible nursery count of eight hundred offspring. Second, if there are no males for an extended period of time, females have the ability to develop a male sexual organ without losing their own ovaries. As such, they can fertilize themselves, thereby giving birth to female offspring. All this is necessary simply because males can't recognize a female. (See also SEX ROLES: ZEBRA FINCHES for how this problem is overcome.)

➾**FYI:** Usually guppies only live two or three years. However, some have lived as long as seven years. Sometimes females who live this old will turn into males.

MARINE WORMS

A popular T-shirt a decade or so ago read, "A woman without a man is like a fish without a bicycle." In the case of many species, this is not satire, but biological fact. This is not a case of cloning either, but actual sex role shifting to permit intercourse. In the case of the marine worm or **nereid** (*Nereis virens*), as long as the animal has fewer than twenty segments to its body, it is a sperm-producing male. But once it grows longer than twenty segments, it begins to produce eggs, not sperm. Which means that all the young in this species are males, while all the larger adults are females. If an adult female has part of her body severed so that she has fewer than twenty segments, she reverts to male again. The drive to mate and produce offspring goes even further: If one were to toss two females into a jar half filled with sand for two or three days, the smaller of the females would metamorphose back into a male in order to fertilize the female's eggs.

OYSTERS

Oysters are among the legion of animals who have the ability to repeatedly change sex. They release their sperm, then over the next few weeks transform into a female and release their eggs. A few days later they go back to being males. One interesting note: All the oysters in an oyster bed change sex at the same time, that time being dictated by the moon.

SEA BASS

Some species of sea bass—including the **giant grouper**, which can grow to thirteen feet and weigh one thousand pounds—remind one of those old sword-and-sandal movies where the mongrel horde swoops into a village to force the eligible young men to join their army. If there was an equivalent fish movie that called for raiding some species of sea bass, they'd be out of luck. They would find no eligible males. Just females and old males.

Where did the young males go? The answer is nowhere. In fact, they never existed. The members are all female until sometime between their fifth and tenth year—then they become males.

SNAILS

There are many examples among animals of females who as they get older transform into males. Some people—mostly males—tend to view this as proof that maleness is the goal that all should strive for, like salvation. But there are some animals that transform from male to female as they get older, among them the snail known as *Achatina achatina*, which becomes a male between the ages of six weeks and a year, but later develops into a female.

TURTLES

Whether a turtle is born male or female depends entirely on the weather. If the temperature where the eggs are incubating is hot, females are born. If the eggs are in a shady nest, usually about ten degrees cooler, then you've got males.

WRASSES

Pity the poor sidekick. He's the one who holds the jacket while the hero displays his swordsmanship, or he waits in the car while the hero woos the damsel. Batman had Robin, Green Hornet had Kato, Zorro had Bernardo. One got the girl and glory; the other got to bask in reflected light. Among the wrasse fish there are two different kinds of males that follow this same pattern: the dominating "primary male," who is large and runs the show, and the smaller sidekick type, who basically acts like a butler. These roles are determined at birth: The primary male is born male; the sidekick is born female and lives that way for several years before transforming into a male. The sidekick is smaller and drabber in color, with three horizontal bands on his side that mark his lower rank, and keep the primary male from pouncing on him.

Their duties as loyal sidekicks include guarding the nest for the primary while he's out at one of his other three or four nests. Sometimes a female may approach the nest while the primary is gone; that's when the sidekick gets his opportunity to prove his maleness.

14
SEX ROLES

Field notes. . .

In *My Fair Lady,* when Professor Higgins becomes frustrated with his inability to fathom Eliza's emotions, he complains, "Why can't a woman be more like a man?" But what makes a man a Man? Or a woman a Woman? These simple questions have no simple answers, though they have remained the quintessential questions for people for thousands of years. How does a man act—that is, what are actions that are uniquely "manly," as differentiated from "womanly"? To what extent are our social roles merely conventions taught to us and to what extent are they biological imperatives? This is more than a philosophical abstract akin to the medieval ponderings over how many angels could dance on the head of a pin. The answer here affects our social structure, our notion of family, even how children should be raised.

Literature often tries to answer these questions, but such answers come in the form of snapshots that when placed together present a montage definition of what our sex roles are. We learn that men prefer violence as a means of solving problems. That women seek men who, after the woman slaps him, forcibly kiss her, which she at first fights and then responds to with passion. Men are logical, moral creatures concerned with spiritual pursuits; women are earthly creatures who must be guided by men less they stray from the path of righteousness. Don't like these definitions? Nevertheless, they are part of the montage we've been teaching our children about their sex roles. And, of course, there are some parallels to this type of behavior in the animal world. On the other hand, there are also animals who treat each other with gentleness, though it is as rare in the animal world as it is among humans.

Even in the animal world, animals often are confused about their sexual identity. In fact, many animals not only can't tell females from males in their own species—they aren't even sure which sex they are. There's a good explanation: In less complex organisms, the genetic information passed on is relatively simple, so there can be a great difference between the appearance and function of the sexes. However, in so-called "higher" species, there is much more complex genetic information being passed on, which reduces the amount of space for sexual traits. It's as if the genetic code for each species was on a computer disk. With "higher" forms, such as humans, there's so much information to pass on, the disk doesn't have room for some of the finer detail work concerning vast differences in sexual traits. Therefore, sexes tend to be much more similar, resulting in some species struggling with sexual identity, their own as well as that of the opposite sex.

This section doesn't answer the debate about what sex roles ought to be, but merely demonstrates that in nature sex roles are often much more ambiguous than we realize. Pounding drums and running naked through the forest doesn't define a male, anymore than breast-feeding and nurturing defines a female. The desire to limit each sex to certain narrowly defined roles may have less to do with nature and more to do with one group maintaining dominance over the other.

AMAZON MOLLIES

What's unique about this fish, closely related to the guppy, is that they are *always* females. And because there are no males, they must breed with males of related species. Once a year they swim outside their community of females for the specific purpose of finding a temporary mate. However, the males from these related species are cannibals who are in the habit of swooping down beneath females who are giving birth, and devouring the young as they come out (a method of birth control not uncommon among some fish). To avoid a similar fate for her own young, the Amazon molly swims away from her own territory to mate with a male. Afterward, she immediately swims back home. Any male who follows, she rams and beats with her fins.

Although the female Amazon molly needs a male to produce offspring, the mating process doesn't work the same way as with most other animals. The male does not fertilize the eggs and thus pass along his own genes. His sperm is more like a cheerleader than a player: It penetrates her egg cells all right, but this just stimulates her own process of cell division. Once this process starts, his sperm fizzles away without ever fusing with an egg nucleus. The resulting offspring are identical to each other—sisters all—and in turn identical to their mother. It's pretty much like photocopying, with each copy identical to the one before it.

➾**FYI: Geckos** are small lizards that have become so popular lately their image appears on numerous products. In Maui, there is a store named Gecko selling nothing but products related to the gecko. Perhaps this popularity comes from their bright colorful skins (they look like some sort of kid's candy). Or maybe it's their unusual ability to reproduce through virgin birth. They are also able to interbreed with different gecko species. Males born of these liaisons are often sterile, but that's okay because the females can reproduce without any help from the males. Their eggs have the same forty-four chromosomes as their body cells, half from their mothers and half from their fathers. The offspring from these virgin births are all females, and they too can give birth without sex. (See also HOMOSEXUALITY: WHIPTAIL LIZARD.)

AMERICAN PHALAROPE

Hormones determine the "manliness" or femininity of a being. Humans use hormonal treatments to adjust chemical imbalances in men who inadvertently grow breasts or women who grow facial hair. The degree of masculinity or femininity in a given individual can usually be traced to their hormonal balance. For this bird, the male and female have a hormonal balance that makes each act in ways we would describe as characteristic of the opposite sex. The female's ovaries produce large quantities of male hormones, which in effect makes the females very "male": They have bright plumage, larger more muscular bodies than the males, and are more aggressive. But the male's pituitary gland manufactures a lot of prolactin, which allows the mammary glands of mammals to nurse their offspring. This flush of "female" hormone makes him docile. That is why among these birds it is the male's responsibility to sit on the eggs and take care of the offspring. The females are polyandrous (have many male sex partners). This social norm developed for the same reason some male birds are polygamous: The sex that develops the bright feathers is the one that gets killed most often by predators. That leaves a shortage of the snappy lookers and a surplus of the drab. Mating rules adjust to the numbers.

➾**FYI**: This aggressive female behavior is also practiced by the **gray phalarope**, the **button quail**, and the **South American painted snipe**. The female **barred button quail** of India is so aggressive that she keeps a large harem of men, and is often captured by Indians and used in their equivalent of our cockfights.

BEE-EATERS

The **zebra finch** learns how to distinguish between males and females by having an abusive father; the imprint of that male abuser is so strong on the male finch that he forever after can identify males as ones that look like Dad (see PARENTING: ZEBRA FINCHES). The bee-eater bird has a bigger problem: Not only can it not distinguish between males and females of its own species, it's not even sure which sex it belongs to itself.

Bee-eaters winter in Africa, but come spring they return to Spain, southern France, and sometimes Germany. Upon their arrival, they begin their search for a mate. The courting process is very passive and sedate: A bee-eater sits on a branch and waits to be joined by another, not unlike a human sitting on a bar stool during happy hour. If a group of birds joins it on its branch, it will chase them away—all except one. This cozy little nook is used only for courtship, not nest building or copulation or anything else. For bee-eaters, courtship requires quiet seclusion because it's so complicated: Not only do they have to determine whether or not they like each other, but they have to figure out which sex the other bird is, and which sex they themselves are.

The problem is that males and females look nearly identical. So to determine their visitor's sex and their own, they take turns playing sex roles. One will act like a male, the other will automatically act like a female. The one playing the part of a female will have to assume the position that allows the male to mount, though the other bird will not attempt actual intercourse—it's only for show, to see how comfortable the role feels. The two birds will continue to switch roles for a while, each playing both male and female many times, until the actual male no longer wants to play the female. That's how they determine what his sex is. And if the other bird keeps making him assume the female role, they then realize he's probably a male, too. When that happens, the courtship process ends right there.

Even when the courtship process determines the two are a male and a female, if neither likes the other's company that much, it proves to them there's no love connection and they separate. This dating game continues for several days until a pair is formed. Even more interesting than their sexual confusion is the fact that, once they have decided which sex they are and have found a mate, they don't engage in sex right away, unlike many other animals. Before mating, they first build their nest on the side of a cliff. The delayed gratification suggests a bond between them that is more than just sexual.

BIGHORN SHEEP

A recent *60 Minutes* segment investigated the extent of rape among male prisoners. An expert announced that such rapes are usually not done by homosexuals, but by men wishing to prove their dominance over others. The bighorn sheep of the Rocky Mountains have a similar rite of passage. Males are generally a contentious lot, their hierarchy based on the size of their horns. The females usually aren't a part of this male bullying; they are more passive and submissive. But for two days a year, when they come into heat, the females act just like the male adults and, boys, you'd better step aside when this happens. They swagger around, starting fights. Such a fight can last for hours, or be over quickly. Either way, the loser crouches in the submissive posture and the victor mounts the loser and has sex. For females, this is the only way they will allow a male to mate with them; he must defeat her in combat. However, the female often defeats the male, in which case she mounts him and they simulate copulation. When two rams fight, the winner has sex with the losing male.

The bighorn sheep is just another in a series of animals that cannot distinguish between male and female. They recognize only winners and losers; the winner has sex with the loser, regardless of what sex it is. It just so happens that the males defeat the females often enough that the species continues.

CICHLIDS

This small fish comes in three distinct sexual roles: (1) males, which are brightly hued; (2) females, which are paler than the male; and (3) males that look and act like females. Though all three types populate each school of cichlid, only a few of the males are sexually active. The other males remain submissive, physically and behaviorally appearing to be females. But when one of the sexually active males dies, one of these passive males rushes in to fill the opening. Like Popeye after downing a can of spinach, his brain activates sex hormones that make him more aggressive and colorful. But if an even tougher male challenges him and he backs down, his scales once again lose color and he returns to his passive behavior.

DRAGONFLIES

Commonly called "darning needles," "devil's darning needles," "horse stingers," or "sewing needles," insects of the order *Odonata* can trace back their ancestry two hundred to three hundred million years. In those days they had a wingspan of four feet, which made them the largest insect to ever live on Earth. Today's dragonfly has a wingspan of four inches. Despite its diminished stature, it is still one of the fastest and most agile of aviators. And it can see everything around it. That's not hyperbole, with eyes containing twenty-eight thousand fixed lenses, the dragonfly is able to see in a 360-degree circumference. Unfortunately, it only spends twelve days in all its glory as a dragonfly before dying. For its first two years it lives as a lowly aquatic grub. With only twelve days in which to live as a dragonfly, it must spend every waking hour eating and breeding.

All this hyperactive breeding and eating takes place in one twelve-foot square bit of territory near a pond or stream. The dragonfly will eat, court, and mate in this small area—in fact, it will live its whole twelve-day life here. During this time, it is the male who takes charge of every aspect of not only the mating, but the actual placing of the eggs. He's like a stage director telling the stage hands where to place the scenery. After sex, he keeps a tight grip on his partner and guides her to a plant he deems appropriate for their offspring. He releases her and she proceeds to lay her eggs on the leaf or stem. The male may then fly off to copulate with another passing female, after which he'll bring her back to this same plant and have her lay her eggs there also, right next to the previous batch. In some species, the male won't release his grip on the female until after she's laid her eggs.

➾**FYI**: Sex for a dragonfly is quite complicated, which may be why the male is so conscientious about making sure the eggs are laid properly. First, his sex organs are on the end of his abdomen. Nearby are two hooks used to hold the female by the neck during sex. The problem is, because he's gripping her by the neck with those hooks, his sex organ can't reach hers. That's why he has an additional set of genitals at the other end of his abdomen. This is the set that will actually engage in copulation. Another problem is that this set—though it consists of a penis, a sac, and a couple more hooks—doesn't contain any semen. That's all at the other end, along with the set of genitals he can't use. That's why the male has to go through a little dating preparation: He curves his abdomen around like a Slinky until the primary sex organ can fill the sac of the secondary genitals with sperm.

HIPPOPOTAMUSES

To many humans, these animals seem so homely as to be lovable and hugable. Perhaps that's why stuffed hippo dolls are so popular. But the real animal can be quite dangerous and unpredictable. Father hippos have been known to literally bite off the head of their own son; sons in captivity have been known to gore their mothers to death. Perhaps it is this dangerous aspect of the male that leads to the females banding together in groups of forty or fifty, mingling with males only for the necessity of mating. (The act of copulation itself usually takes place in the water so that the male can use the buoyancy to displace his great weight and mount the female.)

Knowing all this about the hippo, most people are nevertheless surprised to find that the big lugs are a female-dominated society. Bull hippos are permitted to approach females only when specifically invited through a call. When a male approaches a group of females, he must hunch himself into a posture of humility; when a female rises to her feet, the male must immediately lie down in supplication. He is permitted to rise only after the females have lain down. Any breach of respectful conduct results in the females banding together to drive him away.

➾**FYI 1:** Like hippos, elephants are animals of enormous size and strength, and, as with hippos, elephant society is basically a matriarchy. In fact, elephant herds are made up only of females and their young. The males drift along outside the herd.

➾**FYI 2:** Hippos mark their territory in the same way many other animals do, through the scent of their urine and feces. However, they've added a new spin to the process: While the hippo defecates and urinates, it simultaneously twirls its tail like an electric fan, shooting the concoction in all directions.

LIZARDS

Most guys prefer a somewhat protracted preliminary to an actual fistfight. There's usually some name calling, glowering, chest puffing, steps toward each other, maybe some tentative nudging. But a lot of times the "fight" ends there, before actual blows are exchanged. Maybe it's because each is aware of the damage that can be bestowed due to their size and strength.

The male *Anolis* lizard is pretty similar: When they face off they inflate their throat sacs, raise the combs on their backs, and puff up their bodies so they look three times their normal size. There's also a lot of head bobbing and threatening push-ups. This can go on for an hour until one loses confidence and backs down—a little like a staring contest. Only the loser changes color from bright green to a subdued brown and hurries away. Rarely is there actual physical combat.

Howcvcr, females are a lot less cautious. There's very little posturing; when they're mad they leap at each other and start right in on the biting.

ROCK PARTRIDGES

If ever humans were looking for animals after which to model their utopian society, the rock partridge seems to offer such an example when it comes to equality between the sexes. They form a monogamous marriage that includes shared responsibility for all chores. First, the female digs two nests in the ground at breeding time. Then she lays her eggs in the first nest. The male immediately takes over the domestic duties of the first nest, primarily by sitting on the eggs until they hatch. Meanwhile, the hen scampers over to the second nest, which is a hundred yards away, and lays a few more eggs, which she sits on. These two separate households increase the family's chances for survival; if one nest is attacked by predators, the second may not be. Once the eggs in both nests have hatched, the family comes together in one nest and the next eleven months are spent raising the young. The reason they are able to form this kind of relationship is because they both look similar, neither one more attractive than the other.

SPOTTED SANDPIPERS

At least 90 percent of birds breed in seasonally monogamous pairs, some of which remain "married" for life (see MARRIAGE). But this North American bird is unusual in that not only does it gather a harem, but it is the *female* that gathers a harem. The female is larger than the male and, in another unusual variation, arrives before the males at the breeding ground. Here she competes with other females for the best territory. When the males arrive, she selects a mate, bears a clutch of eggs, and leaves him to care for the eggs and nest while she seeks out another mate. Her first husband may resist her attempt to attract another mate; he may even attack the new object of her affection. But the larger female is quite capable of setting him straight and continuing her search for more studs. And if any eggs are destroyed by predators in any of the nests in her harem, she simply returns to replenish the nest.

WOLVES

In general, wolf society is arranged according to the hierarchy of males, the hierarchy of females, and then by the cross-sexual social hierarchy. An alpha male rules the other males, while the alpha female rules the other females. And, like the football team captain and the head cheerleader, they come together to form the dominant couple—the leaders of the pack. However, there are plenty of instances in the wild of lower-ranking males breeding with alpha females, with no interference from the alpha males.

Females may also lead packs, often outliving the rule of a male alpha. Females typically decide where the pack will den, a crucial decision because it directly relates to how good the hunting is for the next five or six weeks. Young females are slightly faster than young males and therefore make better hunters in some circumstances.

➾**FYI**: According to Barry Holstun Lopez, author of *Of Wolves and Men*, sex roles among humans may have influenced how we perceive wolves: "The male hunter—male leader image of the wolf pack is misleading, but unconsciously, I am sure, it is perpetuated by males, who dominate this field of study . . . I am certain this is part of the reason people believe that male wolves always do the killing, and why males' weights are so often exaggerated." In fact, a wolf may only be alpha temporarily, and only for specific tasks.

15
VIOLENCE

Field notes. . .

Violence among one's own kind generally involves sex—either directly or indirectly. Directly means that two males might sink their teeth into a rival's throat in order to win a lovely female; indirectly means two males might ram each other's skulls over a tiny piece of land because without land they have no chance of attracting a female and therefore of ever mating (see MARRIAGE and PROSTITUTION). Sex is a tool that serves the directive to procreate; violence serves that same master. It is merely a tool to insure individual survival long enough to procreate; although when a tool is used for violence, we call it a weapon. Survival in itself is not an animal's ultimate goal, since no creature survives the inevitable death. Only short-term survival is necessary in order to procreate. That is why some animals are eaten by their mates immediately after sex, or some die naturally after sex. Mission accomplished.

Sometimes among animals, as among humans, stress leads to murder and cannibalism. But with many species it has nothing to do with stress; it is merely expected behavior. Parents eating their young is common behavior in many species of mammals, fish, birds, and insects, both in captivity and in the wild. Sometimes stress will cause this reaction: If the nest or den is approached by some stalking animal, the mother may suddenly turn and slay her offspring, sometimes eating them. This is especially common among birds of prey, gulls, storks, and crows. With mammals, it occurs frequently among carnivores and rodents. But most of the time violence, like sex, is an animal's way of passing along its genes.

ANTS

Americans have always had a fascination with the Japanese kamikaze pilots of World War II. They were trained to fly their planes straight into the enemy—certain death—as a final desperate attempt to serve their country. Some modern terrorists follow the same ideal: strapping explosives to themselves and detonating them when near their enemy. Humans who do such things are usually taught that such a sacrifice will result in their souls ascending directly to heaven. For such a reward, many would gladly give their material lives.

However, the worker ants of a species of *camponotus* from the rain forests of Malaysia perform a similar service to their colony—without any thought of reward. They, too, walk around all day with a bomb attached to their bodies to use against enemies, only the bomb they carry is internal. Inside they have two huge glands that are filled with toxic secretions. When these ants are engaged in combat and things aren't going well for them, they abruptly contract their abdominal muscles, exploding their own body walls and dousing their enemy with poison.

➾**FYI:** Worker **bees** also are capable of the ultimate altruistic sacrifice. When one discovers an intruder bee in the hive, the worker bee will sting this trespasser to death. However, in doing so, its own barbed stinger is torn from its body, killing it as well.

BEES

Immigrants often have a tough time adjusting to their adopted country, especially when native citizens constantly harass and exploit them. We have hundreds of examples in this country, from the deliberate murder of Chinese workers who helped to build the railroad in the Old West, to current sweat-shop practices. The **European honeybee**, imported to Japan about 120 years ago, suffers the same persecution from native **Japanese hornets**. The hornet is three times the size and twenty times the weight of the little worker honeybee, and with his huge jaws, is able to slaughter up to forty bees a minute. A small army of twenty hornets can buzz into a honeybee community of thirty thousand and kill them all within three hours. The honeybees' stinging ability is not enough to kill the giant hornets, so they have devised a killing strategy right out of Edgar Allen Poe: bake the hornets to death.

The Japanese hornet targets a honeybee nest by dabbing the entrance to the nest with its own pheromone, thereby attracting the rest of the gang. This is like the Nazis painting a Star of David on Jewish homes and businesses. When honeybees detect the foreign pheromone on the nest, about one hundred of them go to the entrance and wait. When the hornet comes after them, they duck inside. This is the same strategy Sitting Bull used at Little Bighorn; and, like Custer, the hornet follows the small bunch of bees, knowing he can easily kill them all in a couple minutes. However, once inside the nest, the hornet finds that there are one thousand honeybees lying in wait. Instantly five hundred of the bees swarm over him, creating a bubble of bees around the hornet's body. The bees vibrate the wing muscles inside their thoraxes, which raises the temperature inside the bee bubble to 116 degrees. The bees can survive in temperatures up to 122 degrees, but the hornet will die if the temperature reaches 114 degrees. They continue this baking process for fifteen minutes, enough the kill the hornet.

➾**FYI:** This baking strategy doesn't always work, in which case the bees abandon their nest and build a new one elsewhere.

CHICKENS

If you've ever lost a schoolyard fight, you know how the poor rooster feels after a barnyard battle. It's even worse if you were once the top guy around and you get whooped by some younger new kid. If the highest-ranking rooster in the yard gets his tail feathers kicked by some younger tough, he runs off to the barn and hides in a corner in shame. He will stand there with his head hanging down, face to the wall, perfectly still, not venturing back to the field of shame until the rooster who defeated him is gone. And when he does come out, he will have been demoted in the corporate rooster pecking order.

➾**FYI:** The **rat** is also a poor loser. Though he may not be physically injured in a fight, the mere fact of losing is enough to send some males skulking off to literally die of shame.

FLIES

There are many parasitic animals that lay their eggs on other animals so that the host serves as a living refrigerator for the young to feed off of while they mature. The *Ormia* genus of fly is no different. She deposits her maggots on the much larger cricket and they immediately begin to devour the cricket's muscles. For the next ten days the cricket is eaten alive by his uninvited guests; satiated and fully grown, the fly larvae leave and the cricket dies.

➾**FYI:** This is a familiar story in the animal kingdom. But what makes this one particularly intriguing is that this fly is able to hear differently from the others. Other flies hear with their antennae, which sways under the assault of sound vibrations. But *Ormia* flies have developed eardrums, much like a human's, located at the front of the thorax, just below the head. Why are these flies so different? Because they locate the crickets by the mating sounds they make. In order to be able to hear these crickets, they have had to develop an ear similar to the crickets'. This is a principle called convergent evolution.

FIREFLIES

Female fireflies of the genus *Photuris* aren't just carnivorous, they're diabolically clever. Their method of attracting victims is like something out of a war novel in which the American soldier checks the authenticity of another soldier by asking him who won the World Series. At night the female lies in wait in the grass, watching the male fireflies flitter about. Each of the males of different species has a distinctive signal, which he flashes. He then awaits the return of his signal from a receptive female; when he gets it, he swoops down to join her. The female of the *Photuris* genus has memorized the signals of over a dozen different firefly species. The male, seeing her signal, approaches thinking he will mate with one of his own, but instead runs into this Mata Hari. The surprise doesn't last long because she quickly devours him.

GNUS

Some species of male gnus stake out a section of land and spend every day defending their territory. There's a good reason they're such aggressive landowners: only those males with land get to mate (see MARRIAGE). Those males without land drift around in herds of bachelors until they can stake a claim of their own; otherwise, they will never have sex. This system leaves the bachelors frustrated and the landowners very aggressive and quarrelsome. Each landowning gnu spends his day in border skirmishes with his neighbors.

War between gnus is declared when one gnu deliberately crosses the border onto the property of another gnu. The invader often feigns ignorance, pretending to be so busy grazing as not to have noticed crossing over. However, the defender quickly runs over to remind him of the treaty breach. The two proceed to run around each other in circles, curling their lips at each other, and urinating. Then push comes to shove—literally. They kneel down (so that their horns don't accidentally stab each other in the chest), lock horns, and start shoving each other. After a while they may stand again and start bucking up their hind legs. Finally, the invader returns to his own land. In these border skirmishes the invader never wins and the war is usually over without any real violence. The whole exchange lasts only about fifteen minutes.

Why even bother with this sham war if the outcome is always the same? Because these little wars aren't about actually achieving anything. Gnus don't set out to conquer more land, which they wouldn't be able to defend anyway. They merely wish to vent their pent-up aggression, which they might otherwise turn against the females. If they were this aggressive with the gnu cows, the cows would merely run to the next property owner. In fact, the gnus purposely choose to have land next to other gnus rather than find some isolated unclaimed land elsewhere. Because they live an aggressive lifestyle, the males have to be careful not to let this spill over into domestic violence. Since they cannot repress their aggression, they must drain it away in these mock combats, the way one drains a venomous bite.

➾**FYI**: If the attacker always loses, how do the landless rabble ever get land of their own? A bachelor will invade another bull's territory, but will almost always lose the mock fight and retreat. But he comes back—again and again. His perseverance is his greatest asset—like the seventh member of *The Magnificent Seven*; he was rejected by the group so he followed them until they finally accepted him. Eventually the bachelor will be tolerated among the older bulls and he will have his own territory. But if he is impatient, not willing to wait, then he will have to engage in real combat. Even then, combat is rarely to the death. The bulls lock horns until one is too weary to properly defend himself, at which point he flees.

LIONS

Lions live in a social group known as a pride, which consists of one to four males, several females, and their cubs. They usually confine their hunting to a specific, clearly marked (with urine) area of 15 to 150 square miles. Generally, this is a very stable social environment. However, when a couple new lions come in and either kill or drive off the males of the pride, they are readily accepted by the lionesses as the new rulers of the pride. One of the first acts performed by the new ruler or rulers is the systematic slaughter of all the cubs in the pride. This massacre speeds up the females' coming in heat, at which time the males can mate and produce cubs that bear their own genetic imprint. Part of the reason the lionesses are so solicitous to the new males, despite them having murdered the cubs, is because the stability of a pride allows mothers and daughters and granddaughters to stay with the pride (the sons are generally driven off to avoid inbreeding). New males, with their fresh genetic input, provide stronger offspring with a better chance of survival. (See also PARENTING: LIONS.)

OTTERS

Before you purchase yet another cuddly sea otter doll with the big moon eyes and the hugable baby face, you might want to make sure it's a female, because the males have a nasty habit of kidnapping babies and holding them for food ransom. The kidnapping doesn't occur as often as just direct stealing; one biologist estimates that male otters steal from females as much as one-third of all the food they eat. The typical kidnapping caper usually goes something like this: The young pup otter, who is too young to dive, floats on the surface while Mom does the diving; a male swims over (no one knows whether or not the male is related) and grabs the pup; when Mom surfaces with the sea urchin, the male grabs it, releases the pup, and makes his getaway. The female doesn't put up a fight and the pup is never harmed. Scientists believe males do this because they usually live within small territories that limit their hunting opportunities, while females tend to roam more widely.

SHARKS

Sand sharks and **mackerel sharks** actually produce cannibalistic embryos that battle to the death inside their mother to see which will survive long enough to be born. The pregnant female shark begins with about a dozen small fetuses. But as they grow within her oviduct, they begin to prey on each other. And since they already have a set of daggerlike teeth, it is just a matter of time before the oldest, biggest, and/or strongest, ravenously shreds and gobbles up the others. Finally, as in any gladiatorial arena, there is only one survivor. Fattened on the bodies of his devoured siblings, the lone offspring shifts position in order to trigger its own birth. Now, as it enters the outside world, it is already an experienced hunter, well-fed on its siblings. This unusual form of cannibalism was discovered by a zoologist who was dissecting a dead female shark. While probing the shark's internal organs, he was attacked and bitten on the hand by a nine-inch embryo.

➾**FYI 1:** Siblicide is fairly common behavior. Some species of **snails** also eat their siblings within the pregnant uterus, while in other species the firstborn will eat the remaining eggs. The **Alpine salamander** produces about sixty fertilized eggs, but only one to four will survive the prenatal cannibalism of their siblings. (See also TOADS below.)

➾**FYI 2:** Although sharks are a favorite villain of humans, research indicates that sharks don't really like eating humans. Oh, they'll bite them to death, but they won't eat them. Contrary to their reputation as the garbage cans of the ocean, they don't like to eat lean animals such as birds, otters, sheep, and humans. They prefer animals with lots of fat, like whales, seals, and sea lions. As an experiment, researchers baited sharks with four seal carcasses, one of which had been stripped of its fat. The sharks ate all but the one without fat.

TOADS

Like the shark and salamander mentioned above, the desert **spadefoot toad** endures a harrowing childhood during which sibling feasts on sibling. But with a remarkable difference: The spadefoot produces *two* different types of tadpole, one that is a harmless vegetarian feeding on algae, and one that is a predator with a completely different type jaw and teeth. The cannibals immediately begin to feast on the vegetarians, growing into obese Jabba-the-Hutt-like tadpoles up to seven inches long. The next step depends entirely on the weather. A lot of rain will produce a large amount of algae and the veggie-eaters will thrive. A drought will result in the carnivores having to eat their siblings.

This method allows the survival of the species no matter what the weather.

TURKEYS

Wars used to be family affairs. Monarchies were families that would go to war with other monarch families, sometimes over personal slights to the family rather than great political need. Animals feel such intense family loyalty because a family member carries a similar genetic code. Therefore, survival of family members ensures the survival of that DNA. This concept is very clearly illustrated by **Texas turkeys**.

Hen turkeys are single parents who can lay a couple dozen eggs, hatch them, and raise the offspring all by themselves. But climate and predators will often claim the lives of half a hen's family. Now the mothers join together in a group; when they do, their young start to fight with each other. Because the mothers don't intervene in these battles, those families with more male siblings will generally win. The young females are just as strong as the males, but they aren't as aggressive. The turkeys hatch in April and by December the males separate from their mothers and sisters and form their own groups—literally brotherhoods. These brothers are loyal only to each other; if all die but one, that last surviving male will stay by himself because only brothers by blood can be part of any group. These brotherhoods are like small army units that belong to a larger platoon. That platoon is made up of all the brotherhoods.

Social order within the brotherhood is determined by combat. They peck and claw at each other until one is victorious; he is the leader of the brotherhood. As to the rules of combat, the cocks insist on an honorable fight—*mano a mano*—with no two brothers attacking one. The individual battles can take hours; the loser will stretch out flat on the ground leaving his head and throat vulnerable, should the victor choose to finish him off. This doesn't happen, of course, since brother will not kill brother; they are all allies and will need each other in their fights against the other brotherhoods.

Social order among all the brotherhoods is also established through combat, but this is not as honorable as the family fights. Now brothers will gang up two and three on one in order to defeat the other brotherhoods. The size of the brotherhood is crucial, with groups of at least four brothers usually being the winner. Afterward, each brotherhood has a place on the pecking order. Once the order of ranking has been established through this war, there is no need to ever go through this combat again. Even when the size and strength of individual brotherhoods changes—a brother dies—everyone retains their original ranking.

The brotherhood that rules is the one that kicked the butts of all the other brotherhoods; the godfather of all the turkeys is the brother that has kicked the butts of his bothers within that dominant brotherhood. He is the one who has the best odds of living the longest, since he receives all the food he ever wants. And that's not his only perk. He also gets all the hens. It is only the godfather who can mate; all other males just get to watch. When it's time to mate, the godfather selects his mate and his brothers form a protective circle around him. They and the other males will now look for any predators that might sneak in during the romancing (which takes about four minutes). And if any other cock tries to have sex with a hen, the godfather's brothers hustle over and smack him around. There can be only one breeder—all the other males exist only to allow the Chosen One free access to the females.

➾**FYI:** The group of turkeys described above demonstrate behavior that is specific to certain climates and terrain (see

MARRIAGE: FIELD NOTES); they are from Texas, and dang it, they act like Texas turkeys. But turkeys living in different climates and geographical areas will have different social structures. A species of turkeys living in Georgia's forests do not live in large societies as they do in Texas; the forest terrain doesn't promote it. Rather, they live in small groups. The male turkey gets along with only one or two—three at most—females in his harem. And once the mating is over, so is the marriage. Each wanders off alone.

Oklahoma turkeys start out in societies not unlike those in Texas. They are ruled by the iron waddle of a dictator who promises to keep the grain running on time. But because there is more rain in Oklahoma, the hens here restrict their mating to a specific season; the rest of the time they don't allow the males any sex. When the females are ready to mate, the place goes nuts. The males are so supercharged with desire after the long frustrating sexual drought, they rush to mate, despite the protestations of their leader. It is a bloodless, though not lustless, coup.

INDEX

ABOUT THE AUTHOR

Raymond Obstfeld is an English professor and the author of over thirty-five books. *Kinky Cats* is part of a series of books he has written exploring ethical behavior in humans and animals, which includes *Doing Good: What the World's Religions Teach About Today's Most Controversial Topics* and *SpiritWise: The Moral Teachings of Native Americans*. His writing interests extend to poetry and fiction as well. His novels include *Earth Angel* and *The Joker and the Thief,* and he has a collection of poetry entitled *The Cat with Half a Face*. His mystery novel *Dead Heat* was nominated for an Edgar Award. Indulging his love of music, he also cowrote *JabberRock: The Ultimate Rock-'N'-Roll Quotation Book*.